HUMAN GEOGRAPHY

THE BASICS

Human Geography: The Basics is a concise introduction to the study of the role that humankind plays in shaping the world around us. Whether it's environmental concerns, the cities we live in or the globalization of the economy, these are issues that affect us all. This book introduces these topics and more including:

- global environment issues and development
- cities, firms and regions
- migration, immigration and asylum
- landscape, culture and identity
- travel, mobility and tourism
- agriculture and food.

Featuring an overview of theory, end of chapter summaries, case study boxes, further reading lists and a glossary, this book is the ideal introduction for anybody new to the study of human geography.

Professor Andrew Jones is Head of the School of Geography, Environment and Development Studies at Birkbeck, University of London. Previous publications include *Dictionary of Globalization* and *Globalization: Key Thinkers*.

The Basics

ACTING
BELLA MERLIN

ANTHROPOLOGY
PETER METCALF

ARCHAEOLOGY (SECOND EDITION)
CLIVE GAMBLE

ART HISTORY
GRANT POOKE AND DIANA NEWALL

ARTIFICIAL INTELLIGENCE
KEVIN WARWICK

THE BIBLE
JOHN BARTON

BUDDHISM
CATHY CANTWELL

CONTEMPORARY LITERATURE
SUMAN GUPTA

CRIMINAL LAW
JONATHAN HERRING

CRIMINOLOGY (SECOND EDITION)
SANDRA WALKLATE

DANCE STUDIES
JO BUTTERWORTH

ECONOMICS (SECOND EDITION)
TONY CLEAVER

EDUCATION
KAY WOOD

EVOLUTION
SHERRIE LYONS

EUROPEAN UNION (SECOND EDITION)
ALEX WARLEIGH-LACK

FILM STUDIES
AMY VILLAREJO

FINANCE (SECOND EDITION)
ERIK BANKS

HUMAN GENETICS
RICKI LEWIS

INTERNATIONAL RELATIONS
PETER SUTCH AND JUANITA ELIAS

ISLAM (SECOND EDITION)
COLIN TURNER

JUDAISM
JACOB NEUSNER

LANGUAGE (SECOND EDITION)
R.L. TRASK

LAW
GARY SLAPPER AND DAVID KELLY

LITERARY THEORY (SECOND EDITION)
HANS BERTENS

LOGIC
JC BEALL

MANAGEMENT
MORGEN WITZEL

MARKETING (SECOND EDITION)
KARL MOORE AND NIKETH PAREEK

MEDIA STUDIES
JULIAN MCDOUGALL

OLYMPIC GAMES
ANDY MIAH & BEATRIZ GARCIA

PHILOSOPHY (FOURTH EDITION)
NIGEL WARBURTON

PHYSICAL GEOGRAPHY
JOSEPH HOLDEN

POETRY (SECOND EDITION)
JEFFREY WAINWRIGHT

POLITICS (FOURTH EDITION)
STEPHEN TANSEY AND NIGEL JACKSON

THE QUR'AN
MASSIMO CAMPANINI

RACE AND ETHNICITY
PETER KIVISTO AND PAUL R. CROLL

RELIGION (SECOND EDITION)
MALORY NYE

RELIGION AND SCIENCE
PHILIP CLAYTON

RESEARCH METHODS
NICHOLAS WALLIMAN

ROMAN CATHOLICISM
MICHAEL WALSH

SEMIOTICS (SECOND EDITION)
DANIEL CHANDLER

SHAKESPEARE (THIRD EDITION)
SEAN MCEVOY

SOCIOLOGY
KEN PLUMMER

SPECIAL EDUCATIONAL NEEDS
JANICE WEARMOUTH

TELEVISION STUDIES
TOBY MILLER

TERRORISM
JAMES LUTZ AND BRENDA LUTZ

THEATRE STUDIES
ROBERT LEACH

WORLD HISTORY
PETER N. STEARNS

HUMAN GEOGRAPHY

THE BASICS

andrew jones

LONDON AND NEW YORK

First published 2012
by Routledge
2 Park Square, Milton Park, Abingdon, Oxon OX14 4RN

Simultaneously published in the USA and Canada
by Routledge
711 Third Avenue, New York, NY 10017

Routledge is an imprint of the Taylor & Francis Group, an informa business

British Library Cataloguing in Publication Data
A catalogue record for this book is available from the British Library

Library of Congress Cataloging in Publication Data
Jones, Andrew, 1973-
Human geography: the basics / Andrew Jones.
p. cm. – (The basics)
Includes bibliographical references and index.
1. Human geography. I. Title.
GF41.J65 2012
304.2 – dc23
2011047582

ISBN: 978-0-415-57551-5 (hbk)
ISBN: 978-0-415-57552-2 (pbk)
ISBN: 978-0-203-11800-9 (ebk)

Typeset in Bembo and Scala Sans
by Taylor & Francis Books

CONTENTS

LIST OF ILLUSTRATIONS

TABLES

FIGURES

INTRODUCTION

WHAT IS HUMAN GEOGRAPHY?

The academic subject of geography has had a mixture of fortunes throughout its history. The ancient Greeks saw geographical knowledge as one of the leading forms of scholarship, and the birth of modern geography placed it at the forefront of expanding Western empires in the 18th and 19th centuries. However, geographers were also at the forefront of ideas in a darker phase of history in the 20th century and caught up in ideologies leading to the First and Second World Wars. In the latter part of the 20th century, the subject also lost status. After the Second World War, some questioned the coherence of a subject that spanned the natural science of physical geography and the social science of human geography. Harvard University actually closed its geography department in 1948, more or less for just this reason. Moreover, in the English-speaking world, the later 20th century saw geography lampooned in popular culture as backward-looking, all about the names of capital cities, rivers and drawing maps. By the 1970s, comedians on television and film gained laughs from stereotypes of 'geography teachers', perhaps based on caricatures of teachers boring students with facts about far-flung places. In British culture, BBC comedies such as *The Goodies* and later *Blackadder* (now endlessly repeated on cable channels worldwide) portrayed geography teachers as objects of ridicule. In North America, portrayals have tended more often to be of a dull subject that just wasn't cool.

However, in the 21st century, geography as a whole, and human geography as a part of that, has enjoyed a far-reaching rejuvenation. In the last 30 years, the subject has enjoyed renewed interest and popularity, and influential people beyond the academic world have once again started to echo the 17th century philosopher John Locke (1632–1704) in proclaiming geography to be one of the most useful and important of subjects. Rather than being perceived as a weakness, being *both* a natural and social science is once again increasingly seen as a strength. There are several reasons for this renewal. In part it is to do with a significant evolution of what human geographers study and how they now go about theorizing the social world. It is also to do with how the world has changed, most notably as we live in a world that in the early 21st century in one way or another is increasingly globalized. At the time of writing this book, the current president of the Royal Geographical Society in London is in fact the former *Monty Python* comedian, now famed global traveller, Michael Palin. It is perhaps symbolic of the reversal in the subject's fortunes that such a high-profile figure should invest energy in championing the subject of geography. Undoubtedly, this reversal in geography's fortunes reflects a wider recognition that many of the current and 'big' challenges that face the world today are well addressed by the subject: globalization, climate change, sustainability, economic development or poverty reduction. Yet it is also about a reinvigoration of the theoretical ideas in the half of the discipline that this book deals with: human geography.

In that respect – dealing with half rather than a whole subject – this book is unique in *The Basics* series. To understand geography in its entirety, you may well want to invest in the companion volume *Physical Geography: The Basics*. But human and physical geography are also inextricably linked through geography's long interest in the relation between the social and natural worlds and the ways that many issues – most notably that of our environment – require knowledge and understanding of both.

So what is human geography, and what is all about? Human geography is concerned with all aspects of human society on Earth, but in particular adopts a *spatial approach*. If any one distinguishing feature marks the subject out from other social science subjects, it is this concern to think spatially about the social world. In that respect, human geographers share an interest in an enormous range of topics that are also the concern of other social science disciplines. What makes their perspective different, however, is what many thinkers in the subject call a 'geographical imagination'. Human geographers think about how things exists in space, how features of the social world change across spaces and the difference that places make to the nature of human existence. They are also concerned with the *unevenness* of human existence in space and between different places. This rests on a basic philosophical viewpoint that everything that happens in human life occurs in a certain space and time. Geographers often use the clever epithet that all social life, one way or another, *'takes place'*. That is, everything in human life has to happen *somewhere*, and that *somewhere* (along with its relations to a lot of *somewhere elses*) matters a lot in terms of what actually happens.

Human geography is therefore all about understanding why the spatial nature of 'social things' matter. Differences between places shape how the nature of how things develop. Economic geographers, for example, have long argued that certain industries develop in certain cities or regions for reasons related to the specific nature of those places as well as to their position in relation to *other places*. In previous centuries, iron and steel industries grew up in Western Europe in places that were close to natural deposits of iron ore and in proximity to fuels for smelting like coal. By the 20th century, being close to these raw materials was no longer important but industries persisted in those places because by then a suitably skilled workforce were living in them and other related activities like shipbuilding had started near by. Examples would be north-east England in the UK, or the northern coast of Germany around the Rhine. In the 21st century, however, cheap labour and the demand for steel in developing countries in Asia and elsewhere have increasingly led to the relocation of these industries to new regions of the world such as the southern provinces of China and South Korea.

Likewise, political geographers see the development of certain governments and political institutions in a country as inseparable from the past development of societies in those particular parts of the world. Bolivian politics is very different from Thai politics for a whole myriad of reasons related to the very different locations of these nation-states on the planet's surface, and to the long history and relationships with other places these societies have had. In today's world, where there has been much debate about the globalization of human life on Earth, the patterns of relationships across spaces and places that human geographers have sought to analyse have become increasingly complicated. Equally cultural geographers have long associated the nature of different cultures with – in one way or another – people living in certain places and in certain ways over long periods of time, although in modern times globalization has made this much more complex and difficult.

So human geography then is a very broad subject in terms of topics of analysis but one characterized by a very distinctive emphasis on the nature and significance of space and location. In writing this book I want to try to convince you that it is one of the most useful subjects anyone can study, and that it offers a unique and very powerful approach for understanding the big issues that face everyone on planet Earth in the 21st century. Not to play down the specific strengths of other subjects, human geographers certainly see the world differently from, say, sociologists, economists or political scientists. The philosophical concern with space provides an overall concern with issues that are often dealt with separately in other subjects. This holistic approach to understanding the social world is often seen as a major distinguishing strength. The reason is fairly straightforward: the social world is a complicated and messy thing that requires an understanding of many different aspects in order to see the whole. And you can only get so far in theorizing the world by focusing on one aspect in isolation to the exclusion of others. Economists may focus on markets, political scientists on institutions or sociologists on practices, but human geographers try to look at the relationships between all of these in order to understand what happens in the world. Human geography today is therefore a diverse subject far from the caricatures of geography teachers from earlier decades boring students with factual lists of peoples, places and countries. Hopefully if you are reading this book, your experience of geography in general, and human geography more particularly, is rather different from these caricatures. They exist because it is true that 40 or 50 years ago, the subject of geography was taught rather differently in English-speaking countries, but it is also true that the subject has itself changed quite radically. Before we consider how this has come about, and the sheer diversity of both topics and theoretical ideas in human geography today, it is important to understand the major goals of this book and how it is organized.

THE PURPOSE OF THIS BOOK

This book attempts to provide a whistle-stop tour of human geography to give you a broad overview of the subject. It has deliberately not been organized around a list of what are often called sub-disciplines within the subject. Not only would such an approach be tedious, but covering every possible topic in human geography would be impossible. Other books, such as the *Dictionary of Human Geography* or the online *International Encyclopaedia of Human Geography*, fulfil such a role, and do a good job. Rather the goal of this book is to give the reader an overview that shows how the many different topics and themes in human geography today relate to each other. With that in mind, the book is organized into six further thematic chapters that try to illustrate the linkages between different but often overlapping sub-disciplines in the subject. In that way, it should give you an understanding of how both economic and political geographers are interested in governments and regulations, or how many questions of environmental change concern not only environmental geographers but also cultural or development geographers.

This thematic tour of the subject begins in the next chapter by considering how human geography has been concerned with the big questions around globalization. This debate in human geography is very closely related to the themes considered in Chapter 3: the question of development and debates about the global environment. Chapter 4 then moves to look at how human geographers have conceptualized the states and nationalism, culture and landscape. In Chapter 5, the themes focus on issues that have been of central interest to urban and economic geographers in considering the large body of work within the subject concerned with cities, regions and industries. Chapter 6 then examines themes of a more social and political nature in considering geographical work on population and demography, migration, mobility and labour. This is followed by an overview in Chapter 7 of how social and cultural geographies have sought to theorize the nature of the body and identities based around gender, ethnicity, race and age. Finally, the book ends with a brief concluding chapter that outlines some of the future directions human geography is likely to develop along as a subject.

However, this thematic approach to providing an overview of human geography still does not avoid the necessity of discussing different sub-disciplines altogether. While the thematic chapters do cut across different areas of the subject, these sub-disciplines have distinctive topics of interest and have often developed around particular theoretical and methodological approaches. The major sub-disciplines and the kinds of topics geographers working in them are interested in are shown in Table 1.1. As you will see, human geography is perhaps more interdisciplinary in its nature than other social science subjects, but it is important to realize it is not a chaotic or incoherent diversity. Before we move on to the thematic chapters, it is therefore relevant to consider in more depth the historical evolution of the subject which led to the emergence of these distinct sub-disciplinary areas and also examine the cross-cutting theoretical ideas that are often brought together when human geographers seek to understand the world today. The remainder of this chapter considers each of these issues in turn.

A (VERY) SHORT HISTORY OF HUMAN GEOGRAPHY

While geography as an academic discipline has a very long history dating back to the Greek civilization, the subject we know today emerged during the 18th and particularly 19th centuries as the study of the Earth's physical and human features and how those varied between countries and regions. The development of what is now human geography is in particular bound up in the period when Western European countries were expanding their influence across the globe through the development of first colonies and later empires. Geography as a subject was seen as central to understanding the nature of the world. The first society was founded in 1821 in Paris – the Société Géographique de Paris (SGP), with other national geographical societies such as the Royal Geographical Society (founded in London in 1830) following in European countries soon after. Over the century thereafter, the establishment of geography spread worldwide with, for example, the American National Geographic Society being founded in 1888 and the Association of Japanese

Table 1.1 Human geography and its sub-disciplines

Sub-discipline in human geography	*Examples of topics or debates*	*Example journal/s where work in this area can be found**
Economic	Regional economies Industrial Development Clusters Firms	*Economic Geography* *Journal of Economic Geography* *Regional Studies* *Environment & Planning A*
Social/cultural	Landscape Consumption Identity	*Journal of Cultural Geography* *Environment & Planning D: Society and Space*
Political	International System Nationalism/geopolitics	*Political Geography* *Antipode*
Historical	Past landscapes History of Cities	*Journal of Historical Geography*
Urban	Global urban system Urban development Global cities	*Urban Studies* *International Journal of Urban and Regional Research*
Development	Poverty alleviation Postcolonial Government Post-development	*Journal of Development Studies* *Journal of Latin American Studies*
Environmental	Sustainable development Human impacts of climate change Food security	*Global Environmental Change* *Annals of the Association of American Geographers** *Transactions of the IBG**
Population	Demographic transition Migration	*Annals of the Association of American Geographers*
Feminist/queer	The body	*Journal of Cultural Geography* *Gender, Place and Culture*
Rural	Agricultural change Rural livelihoods	*Rural Geography*
Transport	Mobility	*Journal of Transport Geography* *Mobilities*
Children	Youth identity/exclusion	*Children's Geographies*

* These are just examples, and many journals in human geography span these fields, with some dealing explicitly with a number of areas, such as *Transactions of the IBG*, *Annals of AAG*, *Progress in Human Geography*.

Geographers in 1925. During the 20th century, the legacy of Western European imperialism led to the further spread of geography as a discipline studied and taught in universities across the globe.

Prior to the 20th century, much of what would be described as human geography took a regional emphasis. Human geography in the 19th century was mostly concerned with examining, mapping and describing the distinctive nature of different societies and cultures of people living in different regions of the globe. It has also been criticized for this direct link to the imperial ambitions of Western European countries. Human geography undoubtedly played its part in acting to support **colonialism** and the domination of peoples around the world by Western European societies. Geographical knowledge has always been used by political rulers, military leaders and others, sometimes to ill effect. However, in the 20th century, the problematic status of the subject in relation to the politics of the real world is even clearer. Human geography developed beyond a simple descriptive emphasis providing information about different parts of the world, and began to develop theories of how human societies related to each other in space and territory. Chapter 4 of this book considers, for example, how human geographical theories in the early 20th century were caught up in the world wars. The ideas of the British geographer Halford MacKinder (1861–1947) concerning the competition and conflict for territory between the 19th-century European nations certainly informed political ideologies that led to the First World War. Equally, the work of two German human geographers, Friedrich Ratzel (1844–1904) and Walter Christaller (1893–1969), were made use of by the Nazis in Germany to both justify German territorial expansion and aid the planning of new settlement in countries that had been invaded, such as Poland. It is always hard to judge the past and the intentions of individuals, and these human geographers were not necessarily directly or intentionally involved in the fateful political projects that led to the world wars, but their ideas certainly played a part (Barnes 2011).

Perhaps for this reason the human geography that emerged in the 1950s in Europe and North America moved away from theoretical models and returned to a very regional and descriptive approach. The experience of human geographical theories applied to the real world in the first half of the century had not been a positive one.

However, by the 1960s, human geographers increasingly rejected this regional and descriptive approach, once again seeking to develop a theoretical human geography. Their inspiration was a philosophical school of thought in the social sciences known as **positivism**, which in essence argued that social theories should be developed in the same way as natural science subject such as physics, chemistry and biology. Human geography then took on the methodologies of these subjects, trying to become a *spatial science*. This involved the development and testing of scientific hypotheses, the aim being to establish factual geographical knowledge about the social world and the way it works through the collection of data. During this period human geographers made increasing use of **quantitative methods** and statistical analysis, seen to be more rigorous and scientific than past descriptive approaches to the subject.

Yet by the mid-1970s a further wave of criticism within human geography doubted the capacity of this so-called 'quantitative revolution' to deliver the truth of the social world. In particular, two critical strands to human geography developed. One came in the form of Marxist human geography that argued (drawing on wider Marxism) that attempting to turn human geography into some kind of pure 'spatial science' which could construct objective and neutral facts about the social world was fundamentally misguided. Informed by Marxist **political economy**, which was concerned with social justice, inequality and the uneven power between groups in society, a new wave of human geography rejected the ideas of positivism and sought to develop a human geography that was engaged with political questions. Geographers such as David Harvey and Edward Soja used Marxist theories to offer new insights into why the world economy produced inequalities of wealth in different places. They were in essence interested in the way capitalism led to uneven economic development. Much of this analysis developed a geographical perspective on Marxist theories of capitalism as an economic system that had emerged in Western Europe from the 16th century onwards and spread progressively across the globe through Western European colonialism and subsequent empire-building. Geographers began to argue that this capitalist system – based on money and market relations as the main organizers of economic activity – existed differently and with different effects across and between different places. During the 1980s, Marxist and political economic approaches to human geography became widely adopted as geographers looked at, for example, labour relations in different regions and countries, or how the international production of manufactured goods relies on cheap (and arguably exploited) labour in less developed countries.

The other critical strand was what is known as humanistic geography. This developed from an interest in the philosophical ideas of **humanism** and was critical of the way in which positivism treated human beings as numbers or elements of numerical models. Humanistic geographers argued that the quantitative revolution had 'dehumanized' human geography, and that it no longer paid attention to the context of people in places. Humanistic geography therefore tried to incorporate a consideration of the physical, social and emotional surroundings of people, and moved away from thinking about space as an abstract in the way that the quantitative models of the 1960s had done. This critical line in human geography also saw the use of different **qualitative methodologies** to research the social world which geographers drew from other social science disciplines including psychology, anthropology and sociology.

However, the rise of these critical approaches to human geography soon faced a new challenge. During the later 1980s, human geography underwent a series of further transformations. This is widely described in general terms as the so-called '**cultural turn**' within the subject, and was a result of the impact of **postmodern** and **poststructuralist** philosophies and ideas from outside the discipline. Broadly speaking, during the 1980s, in many social science subjects the idea that social science could produce 'big' generalized theories of the world that explained every aspect of human life came under attack. Postmodern and poststructuralist thinkers argued in various ways that the goal of developing 'big' universal theories of the social world was a mistake and impossible to achieve. In essence, this was for two reasons. First, that knowledge is always a partial and limited thing. You can never fully know everything about something. All knowledge – even scientific knowledge – is a limited representation of reality. This is the case (to differing extents) whether you are dealing with a physicist's theory of subatomic particles, or the kind of social scientific theories found in human geography. The second issue is about the politicized nature of knowledge. Postmodern thinkers argue that there is no objective truth to be found out about the world. It is not only too complex, but the creation of 'truly objective' knowledge is impossible. All knowledge is created by people through a process of negotiation and discussion, and in that sense it is in part always a socially constructed thing rather than something that exists independently. That doesn't mean you can't find out some kind of truth about the world, but it will only ever be part of the story.

There is much more that could be said about postmodern and poststructuralist thinking and it is advisable to read around this issue further (see Further Reading at the end of this chapter). The important thing for our purposes here is to emphasize the substantial impact of a broad group of ideas on human geography in recent decades. Broadly speaking, three effects on the subject are important to highlight. The first is the questioning that postmodern ideas brought to 'big' theories in human geography, whether of the positivist generalizations made by those in the quantitative tradition or the ambitious theories made by political economists and Marxist geographers. Human geography in the last 30 years has been more sceptical about the use of generalized theories, and has sought to develop new kinds of less totalizing ways of creating theory. Second, and related to this, the kinds of concepts that human geographers have made use of have come under critical scrutiny. Human geographers have become doubtful about the capacity of concepts such as class, capitalism, race, gender, society or nature to reveal some kind of general truth of human existence. They now increasingly see these concepts in a more modest light, recognizing the limitations of their relevance. A third aspect of the cultural turn is the diversification in what human geographers have studied and the kinds of theories they develop. As the name implies, the cultural turn in human geography has seen an enormous increase in interest in the cultural aspects of social life, and cultural geographies have proliferated in human geography. It has also addressed issues about the very nature of knowledge, most notably around whether it is possible to 'represent' the world. **Non-representational theories** in human geography challenge the very idea that knowledge should just be a representation of the world, and argue that human geographers need to go beyond representation and think about the practices of knowledge creation (see Chapters 4 and 7).

It is fair to say, however, that the cultural turn has been controversial in the subject, with some human geographers criticizing what they see as the splintering of the subject into many different areas of interest that use different methodologies and are based on different philosophical positions. Others are more optimistic, arguing that this diversification only adds to the power and strength of human geography as a holistic social science. Either way, it seems the cultural turn has opened up a whole range of new dimensions compared to the human geography of 40 years ago.

The history of human geography is therefore at an exciting moment. Distinct schools of thought and approaches in human geography undoubtedly exist and continue to develop in different countries, with human geography in eastern Europe or Scandinavia still retaining differences in approach and emphasis from those in Britain and America. However, more importantly, in the 21st century, human geography is studied and taught across the globe, with the Western European and Anglo-American legacy becoming increasingly diluted. New postcolonial human geographies are becoming much more evident as human geographers in developing countries bring new perspectives and ideas. Latin America, Asian and African human geographers have become more evident, and the subject has reflected on how its past was very much grounded in a Western viewpoint. The history of human geography is therefore an increasingly diverse one, and this book is intended to provide an overview and introduction of many important topics rather than some kind of exhaustive list. Its aim is to act simply as a starting point for readers interested in exploring the breadth of human geography in today's world.

INSIDE HUMAN GEOGRAPHY: KEY CONCEPTS AND THEORIES

We next need to think about some of the major theoretical debates within these different strands to the subject that will appear at various points throughout this book. This could be an enormous task, but the important thing is to have some sense of some of the key theoretical schools of thought and how they relate to social science thinking more generally. In this respect, Table 1.2 tries to provide an introductory overview of the relationships between some of the major sub-disciplines identified in Table 1.1 and relate them both to some of the major theoretical debates within those sub-disciplines

Table 1.2 Theoretical foundations to human geography

Area of human geography	*Theoretical approach or debate*	*Historical key thinkers*	*Examples of related thought in the social sciences*	*Examples of recent geographical thinkers in this area*
Philosophy of social science	Critical realism Actor network theory Structuration theory	Émile Durkheim, Max Weber, Michel Foucault	Anthony Giddens (S) Andrew Sayer (S) Bruno Latour (PhS) John Law (PhS)	Ed Soja
Economic	Political economy Markets/ neoclassical economic theory	Karl Marx, Adam Smith, David Ricardo, Thorstein Veblen	Manuel Castells (S) Paul Krugman (E)	David Harvey, Neil Smith, Brian Berry, Ron Boschma
	New economic geography Finance	Karl Polanyi John Maynard Keynes		Trevor Barnes Gordon Clark, Andrew Leyshon
Regional/industrial	Regional economies Industrial development Clusters/innovation Firms	Alfred Marshall, Nikolai Kondratieff	Michael Porter (IB)	Peter Dicken, Ron Martin, Bjorn Asheim, Allen Scott, Michael Storper, Henry Yeung
Social/cultural	Consumptive geographies Emotional geographies/ affect	Michel Foucault, Jacques Derrida, Sigmund Freud, Martin Heidegger, Gilles Deleuze, Félix Guattari	Zymunt Bauman (S)	Nigel Thrift, Paul Cloke, Sarah Whatmore

Table 1.2 (continued)

Area of human geography	*Theoretical approach or debate*	*Historical key thinkers*	*Examples of related thought in the social sciences*	*Examples of recent geographical thinkers in this area*
Historical cultural	Landscape and representation			Denis Cosgrove, Derek Gregory
Political	World systems theory Critical geopolitics	Karl Marx	Immanuel Wallerstein Benedict Andersen	Peter Taylor, Kevin Cox Gerard O'Tuathail
Development	Post-development	Edward Said	Amartya Sen (D) Arturo Escobar (P)	Rob Potter, Stuart Corbridge
Environmental	Sustainable development		Ulrich Beck (S)	Bill Adams, Yi Fu Tuan Michael Watts
Urban	Global urban system Global cities Urban development	Walter Christaller Ernest Burgess	Mike Davis (S), Saskia Sassen (Pl), Henri Lefebvre (Ph)	Peter Taylor, Doreen Massey, David Ley
Feminist/queer	The body		Judith Butler (Ph) Julia Kristeva (Ph) Iris Marion Young (Ph)	Linda McDowell, Gill Valentine

and with key thinkers both historically and in other social science subjects. The table gives some examples of human geographers who have contributed to these debates in its final column. This should assist in providing a way to think about the way in which some of the major theoretical debates in human geography are situated in the wider social sciences. However, across these debates, we also need to think about some of the key concepts that human geographers use to developing geographical theories of the world.

Foremost are the concepts of space and place already mentioned. With these two, it is important that the significance of time is also highlighted. Much philosophical discussion exists about the interrelationship of space and time in many subjects, and it is well established that you cannot really theorize one without the other. In rejecting the objective ideas that human geography could become some kind of 'spatial science', in the last 40 years the subject has developed an understanding of space as something caught up in and intrinsic to all human practices. For many human geographers, questions about space thus become about something we might call 'relative space', with space defined through relationships between human phenomena. However, there is no single definition that is agreed upon. Rather, different strands of human geography have developed different concepts of spaces, with some considering the physical or material manifestations of space (think of cities or the spatial distribution of industries) while others consider virtual, imagined or symbolic spaces (the way in which we imagine areas of the world such as the 'West' or ideas about the nature of cyberspace). Human geography has thus become interested in a whole range of types of spaces: examples include social, cultural, 'organizational', 'corporate' or the '**embodied spaces**'. Much use is also made in the subject of different categorizations of space, the most obvious of which is the idea of scale. Human geographers often talk about aspects of the social world as existing or applying at a certain scale – the most common of these being tend the 'local', 'regional', 'national' and 'global'. However, the concept of scale is not unproblematic. Debates in the subject have ranged over the usefulness or otherwise of these categories. Where does the local end and the next scale 'up', so to speak, begin? On the other side of the argument, things that are 'global' in the social world are always related in one way or another to many different localities. The global, in that sense, is always made up of things which are local.

This leads us to the way in which human geographers' ideas about space and time give the concept of place a particular importance. Places are usually understood as existing at the scale of the local, even if the nature of that 'localness' is very much influenced by relationships with other distant places and things. Much effort has been made in human geography to think about what places are, and what that means for how we understand the world. It is impossible to review all of those ideas here, but in short human geographers see places as mixture of history, spatial relations and sociality. Everything that human beings do has to happen somewhere. Places are where social relations come together in specific spatial arrangements that last over time, often through what we might call the 'materiality' of the world, and which in turn shape and influence the nature of future social relations. Cities are perhaps the easiest examples of this. Think about the nature of any large capital city in today's world. To understand its significance you need to appreciate a whole range of factors including its long history, the layout and buildings, the culture of the people who live there, and those who pass through. All of these aspects and others are situated together in place, and the nature of many of them is a consequence of that ongoing interaction situated there.

It is important to emphasize, however, that closely related to human geographers' analysis of space are questions of time. Much human geography considers how different spaces develop and change over time, but also human geographers have been very much concerned with the experience of space and time. The two are seen as inseparable, with all human activity existing through both dimensions. A very good example is the widely used idea of **time-space compression** that geographers have made use of to explore how in today's world the world has 'shrunk' or become 'smaller' through globalization (Harvey 1989). Much Marxist geography has been interested in how capitalism has spread across the globe and how economic activity has evolved and changed in different parts of the world. Understanding the unevenness of capitalism across space is clearly impossible without appreciating how these differences have emerged and changed over time. We will consider these issues in more depth in the next chapter.

A second set of concepts that human geography makes a lot of use of are the ideas of systems and structures. The concept of the system is one that has been drawn from the physical sciences – physical geographers of course study the world's atmospheric or climate systems. For human geography, the use of system as an idea by which to understand the world is more difficult. Much social science beyond human geography has also conceived of the world economy as, for example, a global capitalist system or international politics as an international political system. We will consider in Chapter 4 how economic geographers have argued that 'regional innovation systems' are present in some successful regional economies around the world. However, the extent to which human relations are systemic in nature is also the subject of debate in human geography, with some geographical thinking questioning the coherence and qualities of the things being labelled as systems. A similar issue exists around the concept of (social) structures in human geography. Structure is used by some human geographers to refer to enduring characteristics of human societies – Marxist geographers talk about class structures whilst feminist geographers have used the idea of **patriarchal** structures to conceptualize the uneven nature of relations between men and women. Again, the postmodern shift in human geography has questioned whether such things as coherent and consistent social structures exist in the social world, and whether it is useful to think of the social world in such rigid terms when there is a great deal of complexity and dynamism in social relationships.

This relates to a third set of concepts of much concern to human geographers today: agency, power and practice. All of these concepts refer to social action in one way or another. Elsewhere in the social sciences, there is a longstanding debate about the relationship between enduring social structures such as class and the role of individual social actors as individual agents. The key question is the degree to which structures influence or control individuals, or individuals reinforce or change the nature of structures. This structure and agency debate has also been widely discussed by human geographers. More recently, however, in the wake of the cultural turn, geographers have reframed this discussion around questions of how power is understood, and how that relates to what people do in the form of practices. Human geographers have become increasingly interested in understanding the spatialized nature of power and agency that, in many aspects of human life, has been significantly reshaped by globalization. Chapters 2 and 5 consider, for example, how globalization processes have changed the agency and power of nation-states in relation to the growing number of transnational firms and other global actors in today's world.

A fourth important duo of concepts that have increasingly concerned human geographers after the cultural turn is 'knowledge' and the idea of discourse. The former is easy to understand in a commonsense way, but human geography's engagement with postmodern ideas has led to a widespread interest in the politics of knowledge creation. In this context, human geography has become concerned much more with philosophical questions concerning what are known as 'ontology' and 'epistemology'. Put simply, ontology is the theory of what exists in the world. Not all that exists might be knowable, however, and this is where epistemology comes in. Epistemologies are frameworks for knowing what it is possible to know, and hence offer a method for creating knowledge. An important related concept widely used by human geographers is that of a discourse. It is particular associated with the work of the French thinker Michel Foucault (1926–84). The concept refers to certain frameworks of knowledge that have been constructed over time, and how the world is represented using those frameworks (whether that is verbally, in writing or in other ways such as maps). For example, Chapter 4 examines further the complex relationship that various past geographical discourses have with the nature of international politics and world history. The way in which geographical knowledge represented the world in a particular way and how that is bound up with certain sets of power relationships between countries and cultures has been a major focus of much cultural geography. Equally, ideas of discourse run through the discussion of identities in Chapter 7.

Fifth, human geography makes extensive use of the broad concepts of society, economy, culture and nature. Again, there is much debate within the subject as to what these mean and how effective they are at isolating one dimension of the world we live in. Chapters 2 and 5 consider, for example, the various ways in which human geographers have thought about the economy, and the processes that make up economies – most notably those of producing goods or services, and consuming them. However, it also considers how economic geographers in particular have become dissatisfied with the idea that the economy can be understood in isolation from social and cultural aspects of life. Much work in economic geography examines how economic activity is embedded in cultural ideas and social practices within given places or organizations such as firms. Equally, there is an ongoing debate in the subject about the relationship between human culture and the so-called 'natural' world. As several of the chapters in this book will discuss, social, cultural and economic geographers have argued at length that what we mean by nature is a social construction (albeit in different ways). On the one hand, Marxist geographers have argued that nature is produced and that human beings' economic needs are bound up with the physical reality of what we called the natural world. In a different vein of thinking, cultural geographers argue that the category of nature itself is part of human imagination and should therefore be understood as part of culture. No more is this more evident that in debates about what is meant by the idea of 'landscape' considered in Chapter 4.

Finally, a growing body of work within human geography today is concerned with conceptions of the subject, identity, self and other. Again much of this work has come to the fore in the wake of the cultural turn. Human geography in recent decades has been very much concerned to examine what it means to be a human being – a human subject. In this respect, the idea of this human **subjectivity** has been theorized around at least four aspects: the body, the self, the person and identity (Thrift and Pile 1995). These concepts of subjectivity have become important as they provide a critique of the idea that geographical knowledge can be objective, dispassionate and nothing to do with the person creating the knowledge. Much human geography today is therefore concerned about the so-called **positionality** of the writer, researcher or knowledge-creator and how that is bound up with the nature of the knowledge that is created.

Related to this use of the idea of subjectivity, human geographers in many different sub-disciplines – social, cultural, political geography – make use of the concepts of identity, self and other in conceptualizing how people understand who they are and how they differ from others. Chapter 7 explores these concepts in more depth, but regarding identity, the most important thing to know about human geographical understandings of these ideas is that they are multiple and complex. Human geographers argue that it is usually better to talk about multiple identities rather than identity in the singular and that identity is therefore not based on any 'innate' quality of an individual or group but rather exists *in relation* to how we see similarities and differences in others. The concept of the 'other' is important in this respect as it tries to capture how people or things in the abstract are represented as opposite or different to oneself. Much geographical work on **postcolonialism** has used this concept to understand how western cultures were historically understood as superior to those of the East.

FURTHER READING: HUMAN GEOGRAPHY AND ITS CENTRAL IDEAS

There are plenty of textbooks that attempt to cover the breadth of human geography as a subject, but most are caught between the poles of trying to be comprehensive and linking the main overlapping themes in the subject. No single book succeeds fully on both fronts so you are advised to look at more than one. Here are a few of the most useful (in alphabetical as opposed to any order of preference):

Cloke, P., Crang, P. and Goodwin, M. (2012) *Introducing Human Geographies* [3rd Edition]. London: Hodder Arnold.

Does a very good job on conceptual overviews but takes a more specific angle on some of the topics within human geography. You might look at other books to supplement its take on economic geography.

Daniels, P., Bradshaw, M., Shaw, D. and Sidaway, J. (2012) *An Introduction to Human Geography; Issues for the 21st Century*. Harlow: Pearson.

Another comprehensive textbook that takes a more thematic approach to how it covers the topics concerning human geographers with conceptual ideas woven in. This book is stronger on economic, environmental and political geography, and you might look for supplementary reading on cultural topics.

Gregory, D., Johnston, R., Pratt, G. and Watts, M. (eds) (2009) *The Dictionary of Human Geography* [5th edition]. Oxford: Wiley-Blackwell.

This book is generally regarded as an essential resource for students of human geography, and its latest edition is probably as close as it is possible to be to comprehensive in covering the subject within one volume. It is better used as a resource to 'dip' into since the dictionary format is less good at bringing out the linkages between different debates (even though of course the entries are thoroughly cross-referenced).

Hubbard, P., Kitchin, R. and Valentine, G. (eds) (2004) *Key Thinkers on Space and Place*. London: Sage.

An extremely useful book to give you an idea of different perspectives on the central conceptual ideas in human geography. It does so by examining the ideas of a range of different human geographers and other social scientists widely used by geographers.

Nayak, A. and Jeffrey, A. (2011) *Geographical Thought: An Introduction to Ideas in Human Geography*. Harlow: Pearson.

This is an excellent up-to-date and in-depth overview of the conceptual and philosophical debates in human geography today.

Thrift, N. and Kitchin, R. (2009) *International Encyclopaedia of Human Geography*. London: Elsevier.

With 12 volumes and a price tag of more than US$3,000, you won't be picking this up yourself off Amazon, but it is undoubtedly the most comprehensive and detailed resource on the subject. You will need to have access to a library that holds it, though.

2

GLOBALIZATION

This chapter examines how human geography addresses the major issue facing the contemporary world – the concept of globalization. It explores how in contrast to other social science subjects, human geography offers a distinctive approach, with its primary concern for the nature of space and place.

GLOBALIZATION

The history of the last 50 years or so has been a period during which human societies on planet Earth have become more interconnected than ever before. This is what the word 'globalization' means at its broadest level, and the concept is often described as being a 'catch-all' or an 'umbrella' idea because it is used with reference to the increasing interconnectedness across the globe of almost every aspect of human life. Globalization is not just about economic activity (although many people do use it exclusively in that way), but also about all kinds of changes to our existence. That means changes to society, cultures, politics, technologies, the environment and so on. Globalization is therefore about more than the growth of global corporations such as McDonald's or the fact that you can buy iPods everywhere. It is also about the effect of many new aspects to life in today's world – for example, the effect of the emergence of the internet, the massive growth in cheap air travel, the international politics of addressing climate change or the rise of 'global' TV shows you can watch wherever you are on the planet. Globalization is about the emergence (or not) of an integrated human society on Earth.

The word 'globalization' itself is, however, only a recent term for this integration. Its origins go back the 1950s and 1960s with ideas like '**the global village**' and '**spaceship Earth**' adding to the sense of a 'shrinking world'. But it is only since the late 1980s that the concept has been propelled into widespread usage by social scientists in several subjects from management studies to sociology. These days the term is ever present in popular discussion among politicians and in the media, but often it is only used in that narrower sense, to refer to the economic aspects of life.

In contrast to the popular uses of the word, and in common with other social scientists, human geographers have tried to develop a more sophisticated understanding of globalization. They often therefore imagine globalization to be some kind of general process of change, or a set of processes, which are dramatically altering the relationships between people and places, and generating new networks of activity and flows of people, ideas and things across regions and continents. This increasing interconnectedness has of course been going on for a long time historically speaking, as the Roman Empire 2,000 years ago or the Chinese empire (often called the 'Celestial' Empire) in the Middle Ages, both corresponded to earlier but more limited forms of this kind of integration. The important difference, however, is the increasing range, speed and intensity of interconnections that have developed in the last couple of centuries broadly, and the last 50 years most particularly. Since the end of the Second World War, the so-called 'shrinking world' has been shrinking like never before, and the pace of interconnectedness dramatically increased. Geographers and others have come up with several ideas to encapsulate this – the '**annihilation of space by time**', '**time-space convergence**' and '**time-space compression'**. All see globalization as a change in the way in which we experience space and time. This makes globalization an idea very much at home with the heart of human geographers' interests, since in many ways places are where these changes to our experience of time and space come together.

As a subject, human geography is sometimes said to have been late to join the so-called globalization debate. Much of this debate is dominated either by journalists or commentators writing from a policy viewpoint, or by political scientists from the theoretical side. This may be in part because human geographers found early globalization theories rather simplistic, particularly those that famously argued that globalization represented an 'end of geography' because everywhere was increasingly becoming 'the same'. Whatever the reason, in truth the popular globalization debate probably found the more complex approach to globalization from geography hard to grasp. Geographical thinking has often tended to focus on the complexity of the changes brought about by globalization, rather than presenting simpler stories of how the world is becoming interlinked (like the journalist Thomas Friedman's argument that the world is now 'flat') (Friedman 2007). It has also interrogated the uneven and complex way an integrated free market capitalist global economy has emerged over time, and contributed to understandings of how the ongoing development of markets for goods and services in the global economy does not necessarily conform to the idealized models of economists (see box).

ADAM SMITH, THE MARKET AND FREE TRADE

Since the mid-20th century, the logic of economic globalization has been widely associated with the benefits of free market capitalism. In the last 60 years, countries have increasingly organized their economies around markets for goods and services rather than adopting the planned approach towards producing goods and services taken by communist countries. In the early 21st century, almost every country on the planet now has some kind of free market economy (with a few remaining exceptions, such as Cuba and North Korea).

A significant body of social scientific theory (including human geography) argues that free market capitalism leads to maximum benefit in the global economy in terms of both the most efficient allocation of resources and the greatest amount of production. An important basis to these arguments is the thought of the 18th-century political economist Adam Smith (1723–90). Smith famously argued in his book *An Inquiry into the Nature and Cause of the Wealth of Nations* that, in contrast to the highly regulated systems of production and trade at that existed at that time, markets were the best way for people and countries to allocate economic resources. Where markets for goods or services exist, the process of markets operating (people offering things for sale and others buying them) meant that the optimum use of resources was achieved. Smith called this 'the invisible hand of the market'. Importantly, he argued that this applied not only to individuals and groups at a local level but to entire countries and the whole world economy. Smith's ideas remain the underpinning for modern neoclassical economics and the argument that the global economy should have free trade between countries rather than nation-states protecting their economies with restrictions on what can be imported into their territories.

For example, economic geographers focus on the complexity of how manufactured goods are made through complex global production networks that are shaped by many national and regional political contexts and factors. Equally, political geographers have examined the complex politics of global environmental change that involve many actors at different scales from individuals to city governments to transnational organizations such as the UN. This appreciation of complexity certainly gives the subject the advantage of a more sophisticated understanding of globalization, but it has probably hindered the profile of geographers as significant contributors to the wider globalization debate.

THE WORLD SYSTEM

Prior to the recent debate, human geographers had been making use of an earlier idea that in many ways is a forerunner to the concept of globalization. This is the idea that human society on Earth is part of some kind of world system. This approach to theorizing world society has its roots in the **classical social theories** of 19th-century thinkers including those of Karl Marx and Max Weber, which respectively examined the historical emergence of a capitalist system for organizing economic activity (see box on p. 26), and the rise of institutions and organizations in the modern era (roughly since the 17th century). Capitalism is probably the most important concept of a social system used by human geographers. It refers to a form of both economic and social organization where the activities of producing goods and services are separated from both the people who own the means to produce things (capitalists) and those who do the work of production (labour).

MARX'S THEORY OF CAPITALISM

Karl Marx's arguments about the nature of the economic system we know as capitalism remain highly influential theories even well over a century after his death in 1883. Marx was a 19th-century philosopher and political economist who is probably most famous for his political pamphlet *The Communist Manifesto* (1848) written with Friedrich Engels. However, this short book is only really what we would see as a policy document drawn from the substantial body of his scholarly three-volume work on economy and society, *Das Kapital*. We have already met another early key thinker about capitalism – Adam Smith – but Marx's thought represents another important and contrasting set of ideas.

Marx adopted an historical approach to understanding the nature of economic activity, and his work examines how the medieval feudal system in Western Europe based on agriculture and the rule of monarchs evolved into an industrial economic system based on money and private property. Marx argues that the key issue is the way that those with money (capitalists) invest in the production of goods and services by buying land, machines and (importantly) labour. The aim of capitalists is to make a profit by selling goods for more than the total cost of the inputs into their production. Especially significant is that one of the major ways capitalists do this is by paying labour as little as possible.

A further stage in his work theorizes the impact of this economic change on societies. Central is the emergence of different social classes based on those who accumulate wealth (the capitalists) and those offering labour (the working class). For Marx, this capitalist economic system contains a series of problems and contradictions. The major one is that the whole system depends on people buying the goods being made by capitalist industry but, in seeking to increase profits, capitalists have an incentive to pay lower and lower wages. Over time, this means that demand for goods collapses and production is no longer profitable. Capitalists make a loss and sack their workers, making the problem yet worse. This represents a Marxist basis for understanding what we these days call the 'business cycle' of periods of economic growth (boom) and recession (bust).

Geographers have made much use of a whole range of updated Marxist approaches, not all sticking to the historically specific arguments of Marx in the 19th century (which have been widely criticized). One of the uses to which geographers have put Marxist understandings of capitalism is to explain uneven economic development, and the geography of economic growth and crisis. For example, David Harvey has argued that the capitalist system's response to crisis is inherently geographical: when a crisis affects one region of the world's economy, capital investment looks to new places to restore profitability in what he terms 'a spatial fix' (Harvey 2007).

The concept of a system is itself actually a metaphor taken from the natural sciences, where for example the Earth's climate and living organisms are understood as being systems. The degree to which human social life is also composed of things that look like systems is, however, the subject of continued debate. Nevertheless, in human geography the concept of the system is most widely associated with the way it was developed by a social historian called Immanuel Wallerstein (born 1930) who proposed what is known as 'world systems theory' in the 1970s. This is in essence an early kind of globalization theory. It argues that capitalism has become the dominant form of economic organization, spreading across the globe since the 16th century, and that this forms the basis for the modern world system we live in today. Wallerstein suggests that other forms of world system did exist in the past (the Roman and Chinese empires are examples of this), but that they were not as geographically extensive as capitalism has become. Looking at the world map of the late 20th century, Wallerstein argued that the world could be divided into core and peripheral areas. The core corresponded to the wealthier countries (Western Europe, North America, Australia and Japan), and the periphery of the less-developed world (Africa, Latin America and much of Asia). The Marxist aspect to this approach comes from the argument in world system's theory that the core areas derived **surplus value** (i.e. profits) from peripheral ones. Now, while Wallerstein was not a geographer, political geographers have made much use of this theoretical approach and developed ideas of the world system. However, since the rise of globalization as a concept, the idea of the world system has been more broadly used to include, for example, the nature of the relationships between states at the level of international politics. In the ongoing debates within human geography regarding the nature of any world system, an important issue continues to be the degree to which the metaphor of a system is suitable. Much political geography points to the complex and diverse relations between states, regions and different actors in today's world, and it is not altogether clear that these interactions resemble the systems that natural scientists discuss (and in that sense whether society is *systemic* in nature).

GLOBAL SOCIETY

Linked to debates about globalization and the nature of any world system is the idea that a global society has emerged in the last 50 years or so. Social sciences of the 19th and early 20th centuries – including geography – saw societies as largely contained within specific countries, nation-states or regions of the world. The basic assumption was that people lived in geographically restricted communities that marked the boundaries of many different societies across the planet. Since the latter part of the 20th century the argument that has arisen is that these boundaries have broken down or been eroded, and that the wider integration of states, institutions and other activities has produced an integration of state societies into three distinctive blocs. Geographers characterized world society in the second half of the 20th century around a three-way division: a 'First World' composed of the advanced industrial countries (North America, Western Europe, Australia and Japan); a communist 'Second World' (USSR, Eastern Europe, China); and the developing 'Third World' (Central And South America, Africa, non-communist Asia). Subsequently, integration has become much more pronounced with the globalization of recent decades. It is now hard to talk about American or Chinese society as being in any complete way isolated from the rest of human society elsewhere on earth because of various globalization processes.

It is important to remember, however, that this idea of global society does not mean that all human societies have become the same (sometimes termed '**homogeneity**'). People across the globe still live in very different social environments in terms of the organizations they experience, the laws they are governed by and the customs and practices of everyday life. The issue is more the degree to which these differences have been diluted everywhere on Earth by increasing levels of similarity. Human geographers are interested in a range of aspects of this process – for example, analysing how flows of people through migration and travel affect the nature of society and lead to the development of common features across the globe. Geographical thinking has also been concerned with the emergence of what is known as a 'global civil society'. Civil society here is used to refer to the organizations that exist beyond governments, legal institutions and other official bodies. This generally means voluntary and community organizations, charities and other kinds of non-governmental organizations (NGOs). Again, until the last couple of decades, civil society was understood to be part of the wider society existing within individual states. However, globalization processes have led to the internationalization of these groups and activities, hence the idea that civil society has also become global. Famous examples include many campaign groups, such as Greenpeace and Amnesty International or charities such as Oxfam. Much of global civil society is thus formal inasmuch as it exists around various organizations, but the concept also covers other activities (often termed as occurring at the 'grass roots'), which are informal practices and do not have anything to do with a specific organization.

GEOPOLITICS

The concept of geopolitics has had a long and mixed history over the last century or so, and it is notoriously difficult to define since its meaning has changed between periods of history. Nevertheless, it is very much central to human geography and in particular to the sub-discipline of 'political geography'. In fact the origins of geography as a subject have much to do with this concept and with what is also known more broadly as a 'geopolitical tradition' of thought.

In order to understand the importance of this concept it is useful to distinguish between 'traditional' and 'critical' geopolitics. The word 'geopolitics' was supposedly coined by a right-wing Swedish politician, Rudolf Kjellén, in 1899, but it only entered wider circulation after the First World War. What emerged were various forms of traditional geopolitical thinking that sought to develop theories of the power struggles over territory between nation-states. Geopolitics in this traditional sense refers to the way that geographical factors and other 'spatial' relationships shape international politics – this includes things like the rise and fall of nations and why they engage in conflict and war. In other words, geography shapes the nature of politics at the international scale (albeit in terms of a rather simplistic definition of geography in terms of land, rivers, mountains, oceans and natural resources).

While there were many variants to traditional geopolitics, in relation to the history of and current thinking within human geography, at least three aspects are worth highlighting. The first is the influential ideas of the famous British geographer Halford Mackinder (1861–1947). Although Mackinder never actually used the word itself (Sidaway 2008), he sought to develop geography as a subject that would be useful to politicians and others for governing nation-states. He was particularly concerned with a state's security and with threats from one state to another (known as external 'Powers'). In trying to understand how geography affected international politics, his famous contribution was what is known as his 'Heartland thesis' of 1919. This argued that Central Asia was a crucial region in the unfolding political history of the world. Whichever state controlled this territory would, argued Mackinder, have the potential for world domination. Whether this happened or not would be determined by whichever state controlled Eastern Europe (a **pivot area**), and also whether the state or states on the edge of the Heartland (known as **the outer rim**) took preventative action. At that point in time, the implication was that Britain needed to support the creation of a '**buffer zone**' around the Heartland to protect the then globally extensive **British Empire**.

A second key element to traditional geopolitics is 'the organic theory of the state'. The idea here is that any nation-state or country can be understood as being like a living organism. The theory was first laid out by a German geographer, Friedrich Ratzel, and was then later elaborated by another German geographer, Karl Haushofer, in the early years of the 20th century. One of the most important outcomes of this metaphorical link between the idea of a state and an organism was the notion that, like plants and animals, states need space to grow. This 'living space' (or *lebensraum* in German) became notoriously linked to ideas in Nazi (**Fascist**) Germany in the 1930s.

It formed part of the intellectual argument at the time as to why Germany needed to expand in the Second World War. Rather more horrifically, German geopolitics of the 1920s and 1930s combined arguments about *lebensraum* with a geographical understanding of racial/territorial 'purity'. In that sense, traditional geopolitics also contributed important intellectual foundations for the **Holocaust**, in which millions of Jewish people and those of other minority groups were murdered. What is less well known, however, is the wider impact of this kind of traditional geopolitics on other states even after the Second World War. For example, the organic theory of the state is also significant in South American history, where the metaphor of the state as organism was also extended in the 1970s and 1980s. Military dictatorships in Chile and Argentina drew on these kinds of theories to suggest that resistance to their regimes corresponded to subversive 'cancers' that needed to be eliminated (Sidaway 2008).

Finally, no consideration of traditional geopolitics would be complete without mention of 'Cold War era' geopolitics (see box). In the aftermath of the Second World War, the world map was redrawn for more than 40 years along a new political division: superpower rivalry and conflict between the capitalist US-led West and the communist Soviet-led East. In the era of nuclear weapons, fear of **mutually assured destruction** (MAD) shifted international conflict away from direct military confrontation towards other means. Geopolitical thinking proliferated in this world, with for example US foreign policy in the 1950s and 1960s being based on '**containment theory**'. This involved a strategy whereby the US tried to contain the influence of the USSR by encircling it with governments sympathetic to US interests. The worry was that communism would spread like a disease from one country to its neighbours much like a line of dominoes. Fear of this '**domino effect**' produced the policies that led to US involvement in the **Korean** and **Vietnam Wars** as well as its treatment of **communist Cuba**.

THE COLD WAR (1946–1991)

The Cold War began very soon after the end of the Second World War, and was a conflict between the two superpower nation-states that emerged – the capitalist US and the communist USSR. It was so named because the two main states did not engage in direct military conflict (underpinned by a fear of the use of nuclear weapons on each other and their allies). Rather, it was a period of hostile relations between these states, where each feared any increase in the territorial influence of the other. Through the Cold War international security was largely maintained by the superpowers' huge military strength and their dominance in their regions of influence. However, that did not mean that wars did not occur but that they occurred 'by proxy' as each superpower sponsored different sides in various civil wars in regions outside of their territory. Examples include the war on the Korean peninsula during the 1950s, where the USSR and communist China backed the Korean communists and ended up controlling North Korea while the US and its allies held the South. The Vietnam War during the 1960s and early 1970s is another example. During the later Cold War period, aside from these proxy wars and the kind of espionage that forms the plot of so many James Bond films, the superpowers engaged in ongoing competitive weapons proliferation with both states building up huge arsenals of nuclear weapons and conventional armies. By the 1980s, the effect on the USSR's economy of this expenditure was substantial and the overstretching of resources to fund the military build-up is widely considered to be a factor leading to the eventual collapse of the USSR in 1991. With the transition to capitalism of Russia and other post-Soviet states from the early 1990s, the Cold War came to an end.

However, human geography today has increasingly rejected this politically neutral view of geographical factors, arguing that the influence of geography is neither invariable nor timeless. Rather, the nature of geographical influences on politics is specific to historical and cultural circumstances. Such a perspective underpins what is known as the (new) critical geopolitics and has emerged largely in response to the postmodern- and poststructuralist-inspired 'cultural turn' within the subject since the 1980s. Several strands to this critical geopolitics have been influential in the last couple of decades. First, drawing on postmodern ideas, critical geopolitics has been concerned with the language of geopolitics. The key idea here is that of geopolitical discourse. As discussed in the Introduction, the French philosopher and social historian Michel Foucault used the concept of discourse in the 1980s to demonstrate how language does not capture any kind of timeless universal truth about the world, but is more like a framework of meaning that is subjective and politicized. Applying this idea to geopolitics means – in contrast to Mackinder's view – that geopolitics is not simply about describing truths and facts about world society. Rather it is the study of the power relationships and political motivations that produce certain specific understandings of the world political map. These understandings have emerged from very specific cultural contexts and motivations – the Cold War view of the US about the world map and its fight against communism is a good example. Political geographers today are therefore more concerned with how global space is 'written' with meaning, and this kind of critical approach can be applied to any political description of the world.

A second element of critical geopolitics examines how geopolitical practices in the wider world – that is, what politicians, states and other actors 'do' – are important in creating senses of identity that are the basis for modern nation-states. Geographers have argued that national identity is not something that happens naturally but is in fact remade and reshaped as nation-states define who is 'us' and who is 'them'. This activity does not therefore simply reflect differences that already exist between groups of people in places across the world, it also creates them.

Finally, a third important feature of the new critical geopolitics is an interest in popular culture and how the world political map is represented in wider society. Critical geopolitics argues that it is not just the views of presidents and prime ministers that matter in international politics, but the way in which the citizens of states across the planet understand the rest of the world. This varies, and popular culture, like television and film, is a major way in which societies come to imagine the world. The key thing is the interaction between popular geopolitical ideas and the way politicians use these images and narratives that resonate with their citizens. For example, human geographers have pointed to the way that sport metaphors have been an instrumental element in US geopolitical discourses around 20th-century conflict. In the American popular imagination, foreign conflict is justified because the US has to compete 'to stay on top', much as you might talk about a baseball or other kind of sports team trying to get to the top of its league.

GOVERNANCE

In general terms, governance refers to the framework of governing practices that political bodies, organizations or other institutions undertake (such as national parliaments or supranational organizations like the United Nations). The word is a little ambiguous, though, and human geographers tend to use it in more ways than is found in political science. The problem is that it tends to be used in terms of the nature of these organizations themselves and also to describe the relationships between them. Governance then is both about the activity of governing and the collection of actors who go about doing it. What is more, although most commonly used with reference to an everyday understanding of 'politics', human geography also make use of more specific types of governance which are sometimes less obviously about what you might associate with 'politics'. The distinction rests between what you might call capital 'P' Politics – elections, national governments, the dealings of professional politicians and so on – and 'politics' without the capital, which refers to the multitude of everyday political interactions involved in every aspect of human life (within families, workplaces or indeed any organization). Economic geographers, for example, have also become interested in the more specific issue of economic governance at all scales down to individual firms, not just at the level of national economies (which is more like the usage of this concept in economics, management and regional science). The important thing to emphasize, therefore, is that governance means much more than just the activity of elected governments in nation-states. In this respect, we need to consider these different aspects to this idea in human geography more closely.

Concerning 'political governance', the distinction between a geographical and political science interest centres on the issue of **spatiality**. Human geographers are particularly concerned with the spatial nature of the activities of governing and have been critical of the overemphasis on nation-states as 'containers' of governance in other subjects (such as political science and international relations). As mentioned earlier, globalization has had a dramatic impact on the capacity of nation-states to govern their own territories in all kinds of ways. Examples would be the lack of power national governments have over global corporations who make decisions about where to locate factories and hence where jobs are created. Equally, the growth in the number and increasing power of supranational institutions such as the European Union, the **International Criminal Court** and the United Nations means that national governments now have to share governing activity with an ever-growing number of actors that are 'bigger' than states. Equally, in cultural terms, national governments no longer have the capacity to tightly govern national newspapers, television and other media. In this way, the cultural aspects of life are also escaping national level governance. Geography is therefore also interested in the question of how globalization is changing governance in today's world and in the many different kinds of actors that produce 'global governance'. An important idea here is that there is increasingly global governance without were being single world government – that is, the world is still effectively governed but, unlike in past eras, there is no one state or governing power that has oversight over everything. One of the big debates (political) geographers are involved in here is the degree to which this new era of global governance is adequate for tackling the many problems that face global society. For example, geographical thinking has much to say on climate change and whether new attempts at global governance – such as the 1997 **Kyoto Protocol** on greenhouse gas emissions – can tackle the problem of global warming. Another example would be the global financial crisis of 2007–9, with geographers again seeking to understand what kinds of new governance are needed to prevent a financial crisis in one region from spreading across the global economy.

Following on from this, economic geographers have also been interested in governance in more specific ways. Three things that need governing in today's global world are becoming the focus of more and more attention: firms, economies and markets. While the 2007 financial crisis has prompted more work on the last of these, geographers are also grappling with how, for example, ever larger transnational firms such as Microsoft or Nestlé in many ways escape the governing powers of national governments There is also a growing interest in how large global firms govern themselves (known as **corporate governance**), which is no longer so straightforward as it once was, with companies having operations in dozens of countries and employing tens of thousands of people across the globe.

GEO-ECONOMIES

Most people understand an economy to mean the economic activity taking place within a nation-state, but globalization has dramatically changed this. One of the leading economic geographers, Peter Dicken, coined the idea of a 'geo-economy' to describe the geographically uneven nature of economic activity (Dicken 2011). The term could of course be applied to economic activity within nation-states, but Dicken has long been concerned with 'economic globalization', and has, since the early 1980s, mapped the development of globalized production and the internationalization of manufacturing and other industries. Dicken argues that the world economy today should be understood as a complex set of globalized geo-economies. This argument stands in contrast to the traditional view that nation-states each have an economy based and largely contained in the territory they govern. Dicken's point is that globalization has opened up serious questions about what we mean when we refer, for example, to the US, German or Australian economies. He argues that globalization has produced 'a new geo-economy' that is different from previous eras in terms of how the processes of production, consumption and distribution are organized. All three of these processes no longer just happen in a small number of specific places within states, but exist as connections of many activities between places that are linked through flows of material objects (manufactured goods, components) and non-material elements (ideas, knowledge, services). The new global geo-economy is made up of many networks that span the whole globe, with different actors (individual workers, firms, consumers, nation-states) linked into these networks as 'nodes' in different ways.

TRANSNATIONAL CORPORATIONS (TNCS)

Economic geographers have argued that the major actors in the new globalized geo-economy are 'transnational corporations' (TNCs). Unfortunately, this is another ambiguous term because the idea of a TNC is the successor to earlier (and similar) concepts – the multinational corporation (MNC) or multinational enterprise (MNE). These acronyms are all, however, often still used interchangeably elsewhere in social science writing, and care needs to be taken. Essentially, the 'trans-' prefix in TNC intends to imply that large international firms now exist 'across' national economic borders rather than just operating in multiple countries (as the prefix 'multi-' denotes). Economic geographers have charted and mapped the rise of such corporations since the 1980s, but it was really in the 1990s that the term TNC came to be used for some of the largest, most globalized firms. Occasionally, TNCs may also be referred to as 'global corporations' but this concept is used lazily, and any differences between this and the more technical terms 'TNC' or 'MNC' are unclear.

The theoretical basis for distinguishing between a 'multinational', a 'transnational' or even a 'global' firm rests on the degree to which these economic actors are globalized in three dimensions: how they produce goods or services, where they sell them (markets) and how the firm is set up as an organization. Some business commentators started talking about 'global corporations' as early as the 1970s, but in reality these companies only operated in a handful of countries at that time and in many ways just repeated their operations in each country separately. Car-makers such as Ford of General Motors, for example, bought foreign firms like Vauxhall in the UK or Opel in Germany, which made cars in their respective national markets. In other words, multinational firms became multinational either by setting up new, wholly separate operations in another country or by buying up existing foreign firms that already made the same products in another country. Since the 1980s, however, this has changed in several ways. First, there are far more firms operating in many countries and many different industries. While early multinationals tended to be in mineral extraction or manufacturing, service industries such as banking, hospitality (hotel chains), retail and software are all increasingly dominated by transnational firms (see box below). Second, today's transnational firms are not just companies from the rich global North but from many economies. Several of the biggest transnational shipping companies originate from Singapore, for example, and large oil and mineral companies have emerged from Latin America and Australia. Nine of the top ten steel firms in 2010 were Asian. Third, transnational firms these days are set up very differently, with companies organizing many parts of their business at a global rather than a national scale. New product research, finance and advertising are all run at the global level where once each national operation had its own research or finance department. That is what the idea of a shift to a transnational or global corporate organization form is about. However, economic geographers point to the highly variable and uneven way in which this shift has taken place. Some companies now are very much transnational whereas others, despite being very large, are much less so. It varies between firms of different sizes, from different countries of origin and in different industries.

WALMART, CARREFOUR AND TESCO: THE BATTLE OF THE TRANSNATIONAL GROCERY FIRMS

A few years ago a cartoon in a UK newspaper pictured two scientists in NASA. One is asking the other: 'What is our twenty-year strategy?' His colleague replies: 'To get to Mars before Tesco does ... '

Now the British food retailer Tesco may be some way from interplanetary expansion, but as the cartoon suggests, like the other largest transnational food retail firms, its global reach is enormous. In fact, Tesco is only the fourth largest food transnational in the world, the largest being the US firm Walmart and the second largest the French company Carrefour. Both of these companies also operate in at least 15 different countries. However, the transnational nature of their business far exceeds merely the number of countries in which these firms have stores, and the cartoon mentioned above hints at greater power, influence and global capacities. Probably more significant than the tally of countries that these firms operate in is how they organize, manage and exert considerable power over global production networks. Supermarket chains now exert enormous influence over food production and distribution across the globe. If you go into a supermarket in most western countries (and increasingly many less developed countries), you will find a vast array of different products that have been sourced planet-wide. Fresh produce such as fruit and vegetables are often grown specifically to supply supermarket chains with chilled distribution networks ensuring shelves are always stocked. Fresh blueberries bought in Britain or Germany may have been grown by one supplier firm in Guatemala one week and by another in Morocco the next. Companies like Tesco source their thousands of goods worldwide through different buying relationships and sometimes through collaborations with supplier firms.

Transnational food retailing firms are thus in themselves major actors in globalization, shaping economic activity, employment and flows of goods and people between numerous parts of the world. Increasingly more and more of the global population are consuming food produced and distributed through these complex global production and distribution systems that food retailers manage and control. Harder to measure, but equally significant, these transnational food retailers are changing cultural attitudes and food consumption practices worldwide.

An important debate within human geography beyond a narrow economic emphasis concerns the growing power and influence of TNCs (and whether or not that is a good thing). As their numbers and size continue to increase, these very large firms dominate global markets in all sectors of goods and services and they account for an increasing proportion of **total global output**. This has significant impacts on people's lives across the globe. Social and political geographers have, for example, examined the role TNCs have played in deindustrialization (see Chapter 4) and the effect that has had on communities in the older industrial economies as traditional manufacturing jobs have disappeared. Think of the decline of cities like Detroit in the US, or the north-east of England and its steel industry (the remnants of which are now owned by an Asian TNC). Equally, there is the consequence of manufacturing being located by firms in new places where wages are low: many of the clothes and shoes sold by Western high-street retailers are manufactured in factories located in China, Vietnam or Indonesia that have often been criticized for their 'sweatshop' working conditions. Political geographers have also added to debates concerning the eroding ability of nation-states to control economic activity within national territories as investment decisions about where to site production now fall to these corporations. The ability of TNCs to open and close productive operations, along with their ability to avoid regulation and taxes by shifting production to cheaper, less highly taxed and regulated locations, has led critical commentators to argue that they have become too powerful in the context of contemporary globalization. In recent years, they have certainly also become the target for campaigns, boycotts and protests by anti-globalization groups who see them as negative influences on democracy and the distribution of wealth (see the box on 'No Logo' in Chapter 3). Furthermore, such debates have a strong cultural dimension insofar as the goods manufactured and services provided by TNCs shape cultural practices across the globe in complex ways. A burgeoning Chinese and Indian middle class are consuming the same global brands and products as Americans and Europeans, and this is undoubtedly a transformative experience shifting social norms and cultural ideas in those regions of the world.

Overall, excessively positive or negative claims about TNCs should be treated with caution since the term now refers to an increasingly large and diverse number of often very different firms. Human geographers have been critical of how much thinking about globalization is often simplistic and sweeping in its criticism or praise of TNCs and the institutions that are seen to govern their activities in the global economy. Equally, they see the impact on politics and culture of these firms as complex. In reality, there is a growing body of research that points to the complex relationship between TNCs as the key economic actors in the global economy and a range of other actors such as governments, workers, regulatory bodies, institutions and consumers.

GLOBAL PRODUCTION NETWORKS

Related to these arguments about the way in which TNCs organize globalized production is the concept of the global production network (GPN). This is a distinctive concept within economic geography that has sought to overcome some of the rather simplistic ways other social science subjects (and economics in particular) understand production in today's complex global geo-economy. The key issue is that national economies 'can no longer be said to contain production' inasmuch as many manufactured goods 'get made' in multiple places. A product such as a car or even a laptop computer, for example, is likely to have many different components made by different firms at production facilities in many different countries around the globe. Components get shipped from one factory to another, and to make matters even more complicated, other aspects of production – such as design – might take place in yet another set of locations. This makes the labels 'Made in the US' and 'Made in China' both misleading and quite often inaccurate. It also means that it is increasingly difficult to see production as a process that occurs in one given place at a given time. The concept of the GPN therefore aims to provide a better way of understanding the multiple relationships between different firms that are involved in making something. It represents a development of the older idea of a '**global commodity or value chain**', which aimed to capture the way a good or service was made in a sequence with value being added at each stage of the process. GPNs are bigger, more complex networks of many global value chains, and have at least three dimensions that concern geographers: their governance (see above), their spatiality (i.e. how they are geographically distributed) and what is called their territorial embeddedness (the way they are grounded in political and other institutions in specific places).

All GPNs have to operate across a range of scales – the local places where factories are situated, nation-states that have governments, and global markets where they eventually have to sell products across a world with much social and cultural diversity. The important thing to realize, therefore, is that although these are production networks, the consumption of their products is also a key factor because GPNs are ultimately driven by 'the necessity, willingness and ability of customers to acquire and consume products, and to continue doing so' (after Dicken 2011).

GLOBAL TRADE

Trade in the world economy refers simply to the buying and selling of goods and services between actors (individuals, firms, organizations) in different places. As the world economy has become globalized, total trade has grown enormously but trade benefits some localities and not others depending on the nature of their economies. Whilst growth in total world trade stalled during the 2007–9 economic downturn, the long-term trend has been one of enormous expansion. To get some idea of this, in 2008, total world trade measured in terms of goods exported from one country to another amounted to US$15.8 trillion. In the same year, exports of commercial services was worth US$3.7 trillion. Human geographers have long pointed to the unevenness of patterns of trade. Much international trade is concentrated between the wealthier countries in the global economy. However, in today's world this is changing fairly rapidly. In the last decade, developing countries such as China and India have experienced huge trade growth, with China's trade surplus becoming an increasing source of tension in international politics.

Another issue however that human geography is concerned with is how globalization processes have made understanding the idea of trade much harder. The reason is simple: patterns of trade have become more complex and what we might count as 'trade' more difficult to measure. The conventional way was to measure trade at the national level with nation-states counting how many goods and services they exported and imported. However, globalization processes have complicated this in a number of ways. For one thing, a growing proportion of world trade is different parts of the same large transnational firm 'trading' with each other – this undermines the widely held assumption that trade as an activity comes to an end with the consumption of a good or a service. Another issue is the nature of what is traded, with not only services but also new digitized products (such as software, music, film) hard to measure because they are sold and bought in different parts of the global economy.

Human geography is not, however, concerned just with uneven economic patterns of trade. Political geographers are also interested in ideologies that underpin the free trade along with the issue of **trade justice**, along with the political movements that have emerged fighting to make trade across the global economy 'fairer'. The major focus has been the inequalities between the rich countries of the **global North** and developing countries in the **global South**. The poorer countries are often exporters of raw materials and basic agricultural commodities such as tea, coffee, sugar, cocoa and cotton. What is more, much of this production comes from small farmers who rely on one or two crops for their livelihood. Central to this political geography of trade is the way that large TNCs have used their market power to force down prices. Campaigns by charities such as Oxfam and other groups, such as the Jubilee Debt Campaign, have targeted this power, trying to make sure small farmers in poor countries receive a fair price. Such fair trade initiatives are therefore about changing the nature of markets so that they deliver better terms to producers in poor countries

GLOBAL FINANCE

Finance refers to the trade and circulation of many different types of money and financial products. In today's world, both take many different forms. Most money or 'capital' in the global economy exists not as cash but in the abstract as bank loans, mortgages or government bonds. Money fulfils five main functions in any economy: it is the means by which economic things are accounted for (a *unit* of account); the thing by which the value of everything else is measured in terms of (a '*measure* of value'); it *stores* this value; it provides a way of exchanging goods and services (termed a *medium* for exchange), and it is a way of paying for things (a *means* of payment). Much analysis of money and finance is of course carried out by economists and those in other social science disciplines. However (economic) geographers are critical of much of this because they would say it ignored the geographies of money and finance that lie behind what is often called the 'global financial system'.

Geography then takes a more wide-ranging view of finance than other social sciences. Several strands are evident. One is a concern for the social, cultural and political aspects of finance with economic geographers developing a body of work that has looked at how money and finance have become increasingly dominant since the 1980s, as more people work in these industries in many economies, and popular and media attention becomes more focused on finance. Think of films such as *Wall Street* (made in 1987 with a sequel in 2010), or the media fury around investment bankers' salaries in America and Europe after the financial crisis of 2007–9 (see box). Another concern of geographers working on finance has been the relationship between (local) places and the global financial system.

Central to the current focus of human geographers' interest in finance, however, is the globalization of money and, in particular, the emergence of a globalized financial system. What this means in essence is the way that markets for money (which come in many forms in today's world) have become international, and no longer focused on national economies. Since the 1980s, for example, it has become possible for banks and other financial firms to buy and sell many more different currencies, shares and other financial products in an international marketplace with national governments no longer restricted as to how much of their currency (or another country's) can move in and out of their borders. This has been greatly helped by a range of new information and communication technologies (ICT). Where, even in the late 1970s, traders in the City of London or a stock exchange did their financial business face to face in a large hall using slips of paper, now such deals are all done electronically and online.

Geographers have been particularly interested in how finance has extended its influence to every corner of the globe through this integration, and in how more and more economic activity is bound up with financial relationships that extend across the globe. Much attention has of course been focused on the financial markets that trade 24 hours a day across a network of global cities such as Tokyo, New York, London and Hong Kong. However, equally important for geographers is how the besuited financiers on Wall Street or in the Bund in Shanghai are investing in every corner of the globe, integrating rural societies from the interior states of China to the rainforest regions of Brazil into the global capitalist system. Geographers are thus interested in the uneven way is which finance is penetrating the lives of everyone on the planet. This **financialization** process is not however, always considered to be a positive development, with the ongoing integration and growing power of the global financial system argued by some to be creating greater instability and risk in an increasingly interdependent global economy (see box on the 2007–9 financial crisis).

DEBT

One of the most important aspects of the global financial system is debt. What money is owed by whom to whom across the planet has a complicated geography, and the implications of the historical and future development of this geography are the subject of much theorizing and analysis by geographers. Debt of course comes in many more forms than the kinds you may experience in daily life – credit cards, overdrafts or mortgages to buy a house. Debt in the world financial system takes the form of a whole array of financial products that banks and other institutions trade in the markets. These include government and company debt (**bonds**), different kinds of bank loans, shares, as well as a whole array of more complex forms known as '**derivatives**'. For geographers, however, what is important is not so much the technical aspect of the operation of these debt markets (which might interest economists) but rather the way in which debt has affected different people's lives differently across the globe. This happens at a range of scales. For individuals, debt is significant because it shapes the opportunities and constraints they experience in life. If people are heavily indebted they may be unable to raise their **living standards**, have a home to live in or, in many parts of the world, educate their children. However, the geography of national debt (sometimes called '**sovereign debt**') is also important. If nation-states borrow too much, they end up cutting jobs and public services, which can hinder a country's longer-term prospects for economic growth, as well as negatively impacting on the populations that live there. In this respect, development geographers have examined in depth the consequences of events like the so-called 1980s 'debt crisis' when many developing countries were unable to continue to pay back their debts to banks in the developed world. The governments of countries such as Mexico then had to impose huge cuts in their own domestic expenditure, not only leading to hardship among their citizens but also arguably restricting economic growth for many years afterwards. More recently of course in the aftermath of the global economic downturn from 2007, many developed economies – including the US and European states – have struggled with very high levels of sovereign debt.

However, economic geographers have also been concerned with the wider implications of too much debt for the global economy as a whole. The geographer David Harvey, for example, has argued that the increasing power of integrated financial capitalism represents a potentially catastrophic threat to the global economy (Harvey 2011). Debt is a central aspect of this, as too much borrowing leads to financial crises that are no longer restricted to one country or region but are transmitted across the globe through financial markets.

THE 2007–9 GLOBAL FINANCIAL CRISIS AND ITS AFTERMATH

In 2007, global financial markets experienced a crisis that led to the collapse of several of the world's leading investment banks as well as many smaller banks in the global North. The crisis was very much a geographical phenomenon, beginning in North America but quickly spreading to Europe and parts of Asia. In many countries, national governments had to step in and rescue banks that were dangerously close to bankruptcy. A severe global recession followed. Economic geographers have been concerned to understand how this process of spreading is related to the highly globalized nature of global finance in the world today.

The cause of the crisis was excessive lending by banks, particularly in property markets (i.e. mortgages), along with new forms of derivatives that allowed these debts to be turned into commodities traded through the global financial system. It began in the US, where banks had lent large amounts of money to homeowners who increasingly could not afford to repay the debt and whose property was not worth the amount of money owed. The people with the mortgages stopped paying (known as defaulting), and it became clear that the mortgages themselves would not be repaid. Banks across the globe that held large amounts of these mortgages on their books (or the derivatives related to them) had suddenly lost enormous amounts of money. Many posted huge losses (Citibank), while others went bankrupt altogether (the US banks, Lehman Brothers and Bear Stearns). Subsequently, the crisis spread as similar problems with the ownership of overvalued debts ('bad debts') emerged in lending to other regions of the world – for example, the property markets of Ireland, the UK and Spain.

In many countries, national governments decided that some of their largest banks could not be allowed to go bankrupt as it might lead to the full-scale collapse of national financial systems. In the UK, the government in effect bought (or nationalized) three banks (Northern Rock, Halifax and Royal Bank of Scotland) in 2007–8. However, the level of debt taken on by many European countries has continued to create severe problems since that time. In particular, the high levels of sovereign debt taken on by Ireland, Greece, Italy, Spain and Portugal has threatened their membership and even the continued existence of the EU's single currency – the euro. In 2010 and 2011, these national governments continued to need significant loans and support from the European Central Bank and the **International Monetary Fund** to remain within the single currency system.

SUMMARY

In this chapter we have considered:

- Human geographers' extensive engagement with theoretical debates about the nature and significance of globalization in today's world;

- How conceptual debates about globalization are intrinsically geographical in their being grounded in space, time, and place;
- The way in which human geography sees human society and economies as being systemic in nature, particularly in relation to the emergence and ongoing development of a globalized capitalist geo-economy;
- What is meant by the concept of 'geopolitics', and how human geography as a subject has always been concerned with the relationship between politics and territory of governance;
- The nature of global production networks, global trade and the role of an increasingly globalized financial system.

FURTHER READING

Murray, W. (2006) *Geographies of Globalization*. London: Routledge.
Provides a distinctly geographical approach to an understanding of globalization, relating the different aspects of the globalization debate to theoretical themes within geographical thinking.

Dicken, P. (2011) *Global Shift* [6th edition]. London: Sage.
Remains the definitive text by an economic geographer on the nature of the global geo-economy. It is essential reading for anyone wanting to know more about global production networks and the nature of transnational firms in all industries.

WEB RESOURCES

The Global Policy Forum has a wide range of discussion on current debates about globalization: www.globalpolicy.org

Look at the companion site to Peter Dicken's book: www.uk.sagepub.com/dicken6

3

DEVELOPMENT AND ENVIRONMENT

This chapter considers what is meant by the idea of 'development' and how geographers have engaged with the concept. It then moves on to examine the related concept of 'environment', paying particular attention to what environmental problems might be and the nature of environmental politics.

DEVELOPMENT

The concept of development is controversial, and there is much disagreement within and beyond human geography as to what it means, whether it is possible and ultimately whether it is beneficial. In essence it is based on the (widely held) view that certain human societies on planet Earth are more advanced in some way (economically, technologically or even politically) than others. If every country on Earth were considered to be equally advanced, then by definition there would be no need for development. The idea therefore implies some kind of progressive change by which less advanced societies (understood as being within nation-states these days) develop, although there is no universally accepted definition. The word 'development' became used as it is today from the mid-20th century in the aftermath of the Second World War. In a famous speech in 1949, the then US President Truman said that the '**underdeveloped**' world was both a 'handicap and threat to themselves and the more prosperous areas'. The answer was 'modern scientific and technical knowledge' to tackle the impoverishment of these areas of the word. Development was thus about the modernization and economic progress of countries, as measured by increases in the total output of the economies (normally measured using **Gross Domestic Product** or GDP). The goal was for the poor countries of the 'Third World', as they became known, to 'catch-up' with the more advanced and wealthier economies of the capitalist western First World, and to a lesser extent of the communist Second World (these geographical categorizations of the world were discussed in Chapter 2). A strong element to this was the argument that the more developed world needed to intervene and direct the development of poorer countries in order for them to modernize themselves to permit economic growth. This perspective on development became known as the 'modernization school'.

However, by the 1970s, critiques of this idea of development had appeared. For one thing, some critics argued, it was too narrow an idea, focused only on economic factors. It was argued that the concept needed to include a range of different kinds of measure of development, including such factors as the life expectancy of people in a country and how well they were educated. Yet more important was another challenge from development thinkers in the so-called 'less developed countries'. Using Marxist ideas, the Latin American 'dependency school' argued that approaches to development based on capitalism were keeping the poor countries poor, rather than leading to economic growth. These thinkers influenced the world systems theories we met in Chapter 2, and argued that developing countries need to 'uncouple' themselves from the world capitalist economy if they wished to develop, rather than engaging in greater integration. The concept of development thus quickly became embroiled in political and ideological discussions about whether the global economic system (increasingly dominated by capitalism) could reduce poverty and produce progressive change in the poorer regions of the world.

Human geographers were of course very much interested in the intrinsically spatial debates about the nature, effectiveness and ideological basis of development. It should be apparent in light of the earlier discussion of globalization (see Chapter 2) that the two phenomena are entwined. Indeed, one of the most important contributions of geographers is to argue that it is impossible to talk about these processes in isolation. Since the late 1980s, 'development geography' has been very much at the centre of more recent conceptual debates about the very idea of 'development'. At the same time, the modernizing view of development was replaced by an increasingly dominant set of ideas known as 'neoliberalism', which saw an integrated free market global capitalist economy as the pathway to achieving wealth and human progress. Yet the shift towards a broad neoliberal consensus among governments and the institutions of development (notably the **World Bank** and the **International Monetary Fund**) had other effects. One important aspect of this ideology was an increasing removal of government intervention in economies and societies, based on doubt that states could achieve effective development. Instead neoliberal ideas suggested that private investors and firms represented the best actors to maximize welfare in global society. This replaced the idea that states and other organizations needed to lead in developing the poorer regions of the world, with an approach based on foreign investment, firms and the market. In that sense, neoliberal economic globalization was argued to be the way to achieve development.

In the last decades of the 20th century, however, the debate about how development should be achieved became ever fiercer. A range of thinkers (largely from the global South) began to argue that the whole project of modern development was essentially a flawed activity designed to reinforce and maintain the wealth, power and advantage enjoyed by the richer countries. By the 1980s, neoliberal economic globalization was seen as the manifestation of this project, and thus the target for resistance. Before we consider this resistance in terms of events and practices, however, we first need to consider the concept that emerged from this: post-development.

POST-DEVELOPMENT

Since the 1980s, a range of thinkers within development studies and development geography have questioned 'the very idea of development itself'. They argue that (big 'D') Development corresponds to both a concept and practice that originates and is based on the interests of the historically rich and powerful countries of the global North. A key thinker in this 'post-development' perspective is Arturo Escobar, who argues that the post-Second World War view of modernizing development has 'progressively turned into a nightmare' that has 'failed'. Writing in the mid-1990s, he argued that President Truman's project had not only failed in alleviating poverty, but in fact it had created a particular way of both representing the poorer parts of the world and a prescription for how to solve this perceived problem (Escobar 1995). In particular, modern development understood the global South to be full of poverty, disease and ignorance. Its solution was large government-led projects (building roads, dams, hospitals and so on) that were implemented by Western 'experts' and hardly involved the local people who they were meant to help. For Escobar, such projects were ill-conceived, bad for local communities and for the environment. Modernizing development was, therefore, a problem in itself.

In making such arguments, Escobar was drawing on the philosophical basis of a wider shift in the social sciences known as the 'cultural turn' that was discussed in the Introduction. In essence, the questioning of all 'big' theories of the world by postmodern and post-structuralist thinking was brought to bear on ideas about development. Escobar and others also made use of the work of the French thinker Michel Foucault to uncover how the concept of development was laden with a particular set of power relationships and knowledge. Development theories with a capital 'D' were, for much of the 20th century, about the views of rich people in the global North about what *ought to* be done about the poor of the global South rather than whether the economic system (that is, capitalism) that had produced a wealthy global North needed to change. Geographers have become increasingly interested in the latter idea, which is sometimes referred to as 'little d' development, concerned with the geographically uneven and contradictory nature of global capitalism.

This brings us back to our earlier discussion of debates about globalization, which have come to dominate more recent debates within development geography. Since the 1990s, the old geographical categories used by development thinkers have become less relevant – the First, Second and Third Worlds, for example. Some poor countries have achieved significant economic growth, and globalization increasingly means that wealth and poverty co-exist beneath the scale of nation-states. Perhaps most significant, however, is how ideas about whether both development and globalization are a good or bad thing have become entwined. The anti-globalization movement originated in countries of the global South such as Mexico in the 1990s, and the ideological conflicts between neoliberal economic globalization and these resistance movements are in essence the central questions that occupy ongoing debates in the 'post-development era'. We therefore need to consider resistance in more depth.

RESISTANCE

The critiques of development that have evolved over the last 60 years have led to an array of resistance movements across the globe. Human geographers are interested in the geographies of these spaces of resistance and in particular the relationship between resistance movements at different scales in a globalizing world. In that sense, the Marxist critiques of modernization theory that emerged in the 1970s represent an early form of resistance to development that has since developed into multiple ideas and activities. Today it is not possible to understand resistance to development without of course also discussing resistance to globalization. We can, however, offer a brief history of this resistance.

While people did criticize development in the 1960s and 1970s, it was not until the 1980s and the arrival of neoliberalism that resistance appeared in the form of anti-development movements and popular protests in the less developed world. A key moment in this is the Latin American debt crisis in the early 1980s, when governments across the global South cut public services heavily. Many Latin American countries had very little economic growth during the rest of the 1980s, and their populations became increasingly dissatisfied. In some places this led to civil war (Nicaragua) or Marxist-inspired **guerrilla resistance** movements (Colombia, Peru). In the 1990s, however, new forms of popular resistance also appeared. The most famous is the Zapatista movement in the southern (poor) part of Mexico that declared (more symbolic than real) war on the Mexican government in 1993, saying that its (neoliberal) economic policies were doing nothing for the poorest Mexicans. The Zapatistas are famous because they cleverly made use of the internet to turn their campaign into a global one, and they are widely credited with representing the foundation of the global so-called anti-globalization movement.

In this way, by the late 1990s, resistance to development within specific countries had itself become a globalized phenomenon. The anti-globalization movement protested against the broad consensus of neoliberal economic policies that were accepted across the globe. The focus for protests were meetings of world leaders at summits (the **G8** meeting in Seattle in 1999, for instance) and trade negotiations. This anti-globalization movement incorporated a wider range of what are known as 'grassroots' resistance movements from the global North and global South, and also led to a number of offshoot movements like the Jubilee Debt Campaign, which were aimed at getting the rich countries of the global North to write off the debts of the poorer ones. These campaigns have become increasingly popular in nature with famous pop stars and other media figures fronting these campaigns. For example, the lead singer of U2, Bono, was heavily involved in the debt write-off campaign.

Other resistance movements have targeted the activities of transnational corporations, which have increasingly been criticized for exploiting both the people and natural environments of countries in the global South. Examples of such movements include the 'No Logo' campaign (see box) which highlighted the use of low-paid child and 'sweatshop' labour by Western high street retailers to make clothes sold in the rich countries of the North. In this sense, in the 21st century, resistance to development is bound up with a global resistance movement that contests the ability of the global capitalist economy and dominant (neoliberal) ideologies to deliver greater welfare to everyone on the planet. It has also been a global political and cultural phenomenon, existing across the internet and increasingly making use of new social media such as Facebook and Twitter.

THE 'NO LOGO' CAMPAIGN

One of the best-known campaigns aimed at resisting neoliberal economic globalization in Western countries was the 'No Logo' campaign. Building on an internet-based campaign, the Canadian journalist Naomi Klein published a book with this title in 2000. Its major aim was to point to the negative effects of global brands and the transnational corporations that owned them. Klein argued that famous high street brands were supported by unfair and unjust labour condition in the global South, and that transnational firms produced many Western branded consumers goods (for example, clothing and footwear) under exploitative conditions in the global South. Central to this is the argument that workers producing goods were paid very low wages while substantial profits were made by TNCs owning the brands. Klein also attacked the advertising and marketing strategies of these firms, and in particular the way they targeted young people as consumers of the goods (Klein 2000).

Human geographers have been interested in these ideas as they try to understand the nature of globalization and the politics that surrounds today's global economy; Klein's argument shares common ground with the way political geographers have thought about the development of **new social movements**. However, geographical thinking also presents some critical perspectives on the arguments upon which 'No Logo' is based. Foremost is perhaps the simplistic way in which the 'No Logo' campaign presents cultural globalization, and in many ways overstates the power that transnational firms and advertisers have. In thinking about resistance to globalization, human geographers have argued that the interaction between global branding and local cultures is more complex, and that brands have different cultural meanings and values in different regions of the world.

GLOBAL ENVIRONMENTAL ISSUES

As an interdisciplinary social science, human geography finds itself at the centre of debates about the environment. The subject is perhaps uniquely positioned to address 21st-century environmental issues because it brings together economic, political and socio-cultural perspectives to bear, enabling discussion of the many factors that need to be considered when attempting to understand the environment. A further advantage of course is that the wider discipline of geography spans both the social and the natural sciences, and no issue blurs the boundaries between human and physical geographers like the global environment.

ENVIRONMENT AND NATURE

Environment is another of the wide-ranging concepts that gets used in all kinds of sloppy ways by everyone, let alone human geographers and other social scientists. In a literal sense, it is a catch-all term for all the surrounding conditions that influence human beings and the societies in which they live. If you were to live on a space station, then this is a type of environment, but in reality the word is used to refer to the places around us here on Earth. In other words, it is shorthand for the Earth's environment. This obviously includes places that other living creatures apart from humans inhabit, whether this is the remote deserts or deep oceans. The environment encompasses every aspect of these places, whether the living organisms in a place or the non-living elements that it comprises including earth, air and water.

The history of human geography has involved several different ways of thinking about the environment. In the early 20th century, the subject was preoccupied with the now discredited idea of environmental determinism. This viewpoint suggested that the environment imposed tight conditions and boundaries on the nature of human activity in any given place on Earth and produced certain patterns of behaviour. The implication was that societies in the polar regions, for example, evolved very differently from those on islands in the middle of the ocean or in the tropical regions. Since these early theories of the interaction between human societies and their environment, human geography has developed a number of successive ideas including human determinism (where people are able to completely dominate the environment in which they live) and environmental possibilism (where the environment shapes the opportunities that are available to human societies). It is the last of these perspectives that has increasingly dominated the subject since the last decades of the 20th century, but at least equally important is a philosophical shift in how human geographers view the whole idea of the environment, again linked to the shift associated with the cultural turn.

At the heart of this is a conceptual problem that has occupied several social science subjects, not just human geography, neatly encapsulated in the familiar phrase 'the natural environment'. This may seem strange in that, in a commonsense way, most people know what they mean when they think of 'nature'. You might imagine rainforests, jungles, mountain ranges or tropical reefs to be natural and clearly distinct from human landscapes like cities. However, human geographers have shown how 'nature' and 'human society' have blurred boundaries, and that everyday ideas of there being a clear distinction between nature (non-human living and non-living things) and society (human phenomena) is problematic. Increasingly the separation of the two is argued to be 'socially constructed'. This does not mean there is no such thing as 'nature', but rather that it does not exist outside our understandings and representation of the non-human world. Moreover, many things that are regarded as 'natural' (that is, separate from human interference) are in fact the product of a very long period of interaction with human society. Domesticated plants and animals (sheep, pigs, tomatoes, wheat) that humans have cultivated and adapted over thousands of years are very good examples of this.

Environmental geographers have thus in recent times become increasingly interested in exploring the implications of this insight in relation to how, for example, different environments are valued by different groups in society and how power relations shape what is understood to be 'good' or 'bad' changes to the environment. This means that human geography has significant insights to offer on the nature of environmental problems, to which we now turn.

ENVIRONMENTAL PROBLEMS

Human geography's central concern for the interaction between humans and their environment means that the subject is very much about the causes, nature of and potential solutions to environmental problems. What should be clear from the discussion of the idea of 'environment' above, however, is that human geography has become very concerned in its environmental possibilism phase with the power relations that are bound up with the idea. When applied to environmental problems, the insights of the cultural turn have led geographers to focus on the underlying assumptions that define a form of environmental change as either a problem or not. Changes to the environment are happening all the time, and have been throughout the history of the Earth. Many are influenced by humans, some are not, but they can only become a problem if represented as such by society. The idea that an environmental change corresponds to a problem therefore implies **environmental degradation**, but any measure of 'degradation' is based on a pre-existing human view of what a given environment *should* be like.

Also important to human geographers is the scale at which environmental problems are understood to exist and are addressed (or not). It is common to talk of 'global' environmental problems, but geographers are quick to point out that no environmental change impacts uniformly everywhere equally on Earth. Climate change may be a global problem, but different places will experience the consequence of this change to different degrees. This is as much true whether change occurs to the Earth's surface or atmosphere, or to its **ecosystem**. At the regional and local scales, the understanding of and response to natural environmental disasters is also illustrative of this. The then US President George W. Bush was heavily criticized for his response to Hurricane Katrina striking the city of New Orleans in 2005. While the hurricane itself was a 'natural' phenomenon, some scientists argued that such a strong hurricane was related to human-induced climate change. At the city level, many people died in a district where flood defences failed and this provoked a vigorous debate about whether the city and federal governments had invested enough money in adequate flood protection. The environmental catastrophe that Hurricane Katrina produced was consequently a heavily politicized event. Human geographers thus see environmental problems as inseparable from the politics that surround the environment, and in today's world much of this politics also has a 'global' dimension. Before we come to this topic, however, we need to consider a concept that aims to offer a long-term way of tackling environmental problems – sustainable development.

SUSTAINABLE DEVELOPMENT AND SUSTAINABILITY

The need for there to be some kind of sustainable development is based on the idea that the Earth's resources are finite and that current forms of human activity cannot continue depleting them irreversibly. If everyone on Earth continues to act in their own self-interest, the shared resource it represents will be depleted in a way that is against everyone's long-term interest (this is known as 'the tragedy of the commons'). Unsustainable human activity therefore needs to be replaced by sustainable development, the most widely used definition of which comes from the UN's Brundtland Commission of 1987, which defined it as 'development that meets the needs of the present without compromising the ability of future generations to meet their own needs'. However, considerable debate exists regarding this definition within and beyond geography.

Before the Brundtland Commission, environmental thinking tended to see the relationship between economic growth and the environment as oppositional – growth would be 'bad' for the environment. Central to this is what is known as a 'deep green' approach to development, founded on ecology. In essence, the 'deep green' view of development rejects the idea that the Earth's environment can be maintained without limiting economic growth. One of the most well-known proponents of this is the writer James Lovelock who proposed his 'Gaia' idea in the 1960s – that the Earth is like a living organism and needs to be maintained in good health. Lovelock and other deep green thinkers generally take their ideas from ecology and theories of how ecosystems are maintained or are damaged. Deep green arguments therefore advocate a dramatic change in the way we all live on Earth, suggesting people should make and consume less, produce less waste, travel less and be aware of the impact on the environment in everyday life. In contrast, the Brundtland Commission had a very different view: that growth and the environment could complement each other. This perspective sees sustainable development as ecological modernization and, since the 1980s, has become the dominant way in which governments and international institutions such as the UN and World Bank refer to the idea. Sustainable development from this perspective is about maintaining an environment that serves the economic needs of the future, rather than preserving or protecting existing environments (whether that is species such as the blue whale or habitats such as the Amazon rainforest) for their own sake. The sustainability imagined in ecological modernization therefore does not necessarily mean sustaining an environment that has not been changed (or arguably degraded).

A consequence of this is that 'mainstream' sustainable development cast as ecological modernization has forced many radical environmentalists to disown the concept. Critics doubt above all that ecological modernization is producing or is in future likely to produce 'truly' sustainable development. They point to continuing (and accelerating) environmental degradation across the globe, as well as to a wide range of activities perpetrated by governments, TNCs and other actors, that are really unsustainable. This critique suggests that the early 21st century global economy continues to remained focused on growth at the expense of the environment and that as it grows, an ever-larger proportion of the Earth's natural resources are being exploited. Human and environmental geography today therefore has a central interest in evaluating these competing claims about sustainability. They have also equally become interested in the global politics that surrounds the question of sustainable development.

GLOBAL ENVIRONMENTAL POLITICS

The debate about whether (and how) sustainable development can be achieved at the planetary scale is not just an academic question, but has become a central aspect of international politics. International political concern about the environment began to take shape in the 1950s and 1960s, but the most important moment was the UN summit held in 1972 in Stockholm, Sweden. The Stockholm conference achieved little in terms of actual international agreements or laws, but what it did do was push environmental issues onto the international political agenda. Before that point, meetings of world leaders rarely addressed environmental issues. During the 1970s and 1980s, the environment became an increasingly pressing political issue and countries began to negotiate and come to agreements, as well as take action. In part politicians in the global North were responding to the growing concern about the environment among the populations they represented, but also a string of environmental problems that crossed international borders were identified. Examples include the depletion of the Earth's ozone layer by chemicals commonly used by industry and in consumer goods (chlorofluorocarbons or CFCs) and the political response that was needed across Europe and beyond to the Chernobyl nuclear accident when a Soviet (Russian) nuclear power station blew up sending radiation over a huge area in 1986.

However, since the 1990s, global environmental politics has moved from being a side issue at meetings of world leaders to one of the most important topics. What has produced this change of course was not the 1987 Brundtland Report, or even the Rio 'Earth Summit' in 1992 (although both were important), but a growing body of scientific evidence that human beings are causing a warming of the planet. Global warming has in that sense created an urgent need for political institutions to come to formal agreements that straddle the globe in order to tackle the problem. The first major step in this direction was the 1997 Kyoto Treaty, where a number of wealthier countries signed up to an agreement to reduce emissions of the **greenhouse gases** that cause global warming. However, since the late 1990s, the scientific evidence has shown this to be far from sufficient, and there have been successive meetings of world leaders negotiating further controls of emissions. This has been a major test of whether or not the international political system has the ability to effectively agree on and address global environmental problems. To date, most commentators would agree that these political attempts to tackle human-caused climate change have met with only limited success.

Human geography is especially interested in what can be termed the 're-scaling' of environmental politics to the global level that has occurred in the last 50 years or so. As with other aspects of political globalization, environmental politics in today's world is practised through a large and growing number of actors that exist at many different scales. And, as has already been mentioned, the 'global-ness' of the environment is not unproblematic. The nature of global environmental politics is all about the impacts of current or future environmental changes, such as human-caused climate change, on many different specific places across the planet, where different groups of people and different institutions (nation-states or super-states such as the EU) have interests. The strength of human geography over other subjects in understanding the politics of the global environment lies in its capacity to theorize how all these multiple scales interrelate to shape the actions, agreements and forms of governance that emerge (or fail to).

SUMMARY

In this chapter we have considered:

- What is meant by the concepts of 'development', 'post-development' and debates from a political geographic perspective on different forms of resistance to development that have emerged;
- How human geographers have engaged with debates about the global environment, particularly in relation to the utility of a geographical approach to understanding what a global environmental problem is and how it may be tackled;

- Debates about what 'sustainable development' means and how human geography offers a strong approach for understanding the nature of global environmental politics in the 21st century.

FURTHER READING

Adams, W. (2008) *Green Development: Environment and Sustainability in the Developing World* [3rd edition]. London: Routledge.
Represents one of the most comprehensive accounts by a geographer on the issues surrounding sustainable development.

Castree, N. (2007) *Nature – Key Ideas in Geography*. London: Routledge.
Gives a short and focused account of the major approaches to nature within human geographic thought.

Painter, J. and Jeffrey, A. (2010) *Political Geography: An Introduction on Space and Power*. London: Sage.
Of the many political geography textbooks, this book is particularly successful at applying recent conceptual debates within human geography to current case studies.

Willis, K. (2011) *Theories and Practices of Development* [2nd edition]. London: Routledge.
This book provides comprehensive coverage of development geography.

WEB RESOURCES

The UK government's supported organization aimed at promoting sustainable development is a good place to get an idea of how this issue shapes the policy of nation-states: www.sd-commission.org.uk

The World Bank's You Think! site is a basic guide to the major issues of development: http://youthink.worldbank.org. When you are done with that, try the US-based think-thank Center for Social Development site www.cgdev.org

4

STATES, NATIONS AND CULTURE

This chapter examines the way human geographers have understood the interrelated evolution of states, nations and nationalism. It considers how the modern nation-states that cover the planet today are the consequence of a long and complicated interaction between history, political struggle, different cultures and shared identities – all of which are strongly influenced by geographical factors. The chapter then moves on to consider the related issue of how geographers understand the difficult concept of 'culture' more generally, paying particular attention to its relationship to place and how we consume space. This leads neatly on to a discussion of human geographers' arguments about the nature of landscape.

STATES AND NATIONALISM

One of the commonest confusions in political geography comes with the use of concepts of 'state' and 'nation'. They are often coupled together as 'nation-state', but in popular discussions and the media you will often hear all three of these words used quite interchangeably. They are of course all interrelated, but it is important to understand the differences between their respective definitions and their relationship to the phenomenon of 'nationalism'.

STATES

The 'state' refers to any governing institution that has jurisdiction over a piece of (land) territory on the Earth's surface. A state is therefore an institution that governs a community of people who live in that piece of territory, usually involving some form of social hierarchy with an elite group at the top of it. The idea of the state is therefore fairly old, certainly dating back several thousand years. In discussions of the founding ideas about what a state is and why we need it, it is often the works of classical Greek philosophers such as Aristotle and Plato that are referred to. The ancient Greek civilization was composed of a range of 'city-states' that governed themselves and an area of territory around them. Of course the Roman Empire that followed corresponded to a much more geographically extensive state-like institution (although the idea of an empire has some distinctive features). In more recent centuries, city-states existed and small countries (kingdoms if you like) existed in a patchwork across Western Europe. Think of the world Shakespeare describes in many of his plays set in what is modern Italy. Plays such as *Romeo and Juliet* take place in city-states where the 'state' corresponds to the court of a duke or a local lord. Another good example is the fairy tales of Hans Christian Andersen. These are essentially 19th-century romanticized stories about a medieval European world of small 'kingdoms' that formed a patchwork of small states across Western Europe throughout the Middle Ages and early modern period. In our popular culture, all these ideas are mixed up, and of course often sanitized in children's stories; all those large castles with towers and moats in central Europe were built that way because the medieval world was a violent one where small states were often in a state of conflict and war with each other.

The important historical change that led to the appearance of the kinds of states that cover maps in today's world came with the development of **modernity** from the 16th and 17th centuries. This term corresponds to a number of changes to society linked to the re-emergence of science as an important form of knowledge along with the spread of capitalism. A further aspect was the circulation of many ideas dating back to antiquity, and in particular the writings of Greek philosophers about what a state should be (a body that represents and is answerable to its citizens based on democracy). This challenged the right of European monarchs to rule. Such a process took many centuries and is historically complex, but together these changes produced the progressive transformation of the medieval monarchy-based states into modern nation-states. Like historians, historical geographers are interested in the diversity of ways in which this transformation occurred, but the important outcome was the major shift in how collective identities of people changed.

States therefore are the political institutions that control a certain piece of territory. Even at the beginning of the 20th century, imperial states with borders that were not clearly defined governed much of the Earth's land surface. Examples include the Austro-Hungarian Empire, which covered much of central Europe, and the Ottoman Empire in the eastern Mediterranean (what is now Turkey), across the Middle East to Iran and Iraq, as well as parts of Russia and the Chinese Ch'ing Empire. The Austro-Hungarian Empire had some characteristics in common with modern nation-states, but the others were not defined by the shared idea of being a nation. Rather, different structures and ideas explain why these empires held together, including religion and ethnic or linguistic commonalities.

This territorial control that states possess is encapsulated in the idea of sovereignty. The use of the word, of course, has historical roots in the power of kings and queens ('sovereign' monarchs) as the heads of state. In the modern era, this power has shifted from being that held by a monarch to the power of a state as an institution to govern its territory. Sovereignty has, however, many different forms and, importantly, does not necessarily have a close link with territory in history. In the medieval period, for example, the territories of states were often ill-defined and the borders unclear. The English 'state' that Shakespeare's Henry V talks about included patches of what is now western France and other small areas of other European territories. Sovereignty in this period was also shared with various institutions including the Church and local lords, knights, dukes and so on. Elsewhere in the pre-modern world, forms of political sovereignty were also complicated arrangements with loyalty to religious leaders often being the primary factor in deciding who governed whom and where. The point is that until the ideology of modern nationalism developed in the 19th century, the relationship between states, territory and sovereignty (that is, political power and control) was complex and fluid. It was only when nationalist ideology spread across the globe during the 20th century that the nation-state became the dominant form.

NATIONS AND NATIONALISM

The concept of a nation refers to a community of people who share a common identity, based on some degree of cultural commonality; it often (but certainly not always) entails a common language, and some degree of common ethnic heritage. It is closely related to nationalism, which in short is the ideology that has driven the creation of modern nation-states and remains crucial to their continued existence. If the nation-state is the institutional form of states in today's world, then nationalism is the idea and value-system that underpins those institutions. In seeking to understanding the complex and varied relationship between nations, nationalism and nation-states, human geographers have drawn heavily on the widely, cited arguments of Benedict Anderson that were developed in his 1983 book *Imagined Communities: Reflections on the Origin and Spread of Nationalism.*

Anderson argues that nations are dynamic communities of shared identity that must in essence involve some kind of spatial imagination. There are three (main) aspects to this imaging process. First, nations are imagined as being 'limited' inasmuch as they must have external boundaries beyond which other nations exist. Nationalism as an ideology requires that there are other people out there who are not part of your nation, and nations are defined as much by the people who are not part of them as by those who are.

The second aspect to the way in which nations are imagined brings us back to the idea of sovereignty. Nations have to be sovereign, says Anderson, because the idea came out of that historical period associated with the Enlightenment in the 17th and 18th centuries. The (first) **French Revolution** in 1789 was all about destroying the legitimacy of kings and queens 'appointed by God', and breaking free. The sovereign state is the emblem of this freedom. Finally, nations are imagined as a community of equals. No matter how much inequality and exploitation may go on within nations, the idea is based around a common sense of comradeship. People are loyal to nations and to their national 'brothers' and 'sisters'.

While other subjects such as history and politics have enormous interest in nation-states, human geographers (and most particularly political geographers) have a particular perspective since all three of these dimensions to the imagined basis for nations have a very strong link to territories and to particular places. Nation-states have geographies which are mapped and (re)presented to the national population. At school, everyone encounters this in the modern world as atlases, and geography lessons that teach children in a nation-state about its history and its geographical form. The important point is that all of this is a very recent development. In the 19th and 20th centuries, nationalism provided the basis for a series of practices that in essence created the idea that nation-states should exist.

Nation-states thus rely on what have been termed 'national myths' about the legitimacy and supposed naturalness of these imagined communities. The examples are numerous, but consider how every nation-state has a national flag, national monuments, national museums and national public holidays. National myths often perpetuate the idea that imagined communities are very old, based on the idea that nations have a long history in a given place, when in fact they are much more recent. Their goal is to generate among populations strong feelings of belonging to the nation. Whether that is English nationalists celebrating on St George's Day (St George being the patron saint of England from the early Middle Ages) or Indian nationalists linking a Hindu national history to the whole of the Indian subcontinent (see box), this represents the tendency in nationalism to 'reinvent the past'.

INDIAN NATIONALISM

The modern nation-state of India came into existence on 15 August 1947, after a long nationalist struggle against the British Empire in the first half of the 20th century. This nationalist movement involved several groups fighting for a unified nation-state covering the whole subcontinent. Central to this was the role of Mahatma Gandhi who, as a key member of the Indian Nationalist Congress (INC), led a campaign for independence from the British empire. However, the Indian nationalist imagination was not as coherent as Gandhi had hoped, and the new country quickly divided in two along religious lines between Hindu and Muslim. This 'partition' produced the new nation-state of Pakistan, but also led to violence and the death of an estimated million people (Chatterji 2007). Furthermore, even since Partition Gandhi's imagined community of 'one India' has always been fractured and contested, with regional separatist groups fighting for independent states in Jammu and Kashmir in the north-west. Indian nationalism in the 21st century has several variants, with Hindu nationalists imagining India as a Hindu-only nation in contrast to the secular vision of India as a mixed religious community of multiple religions and ethnicities that has dominated for most of its history. If you are interested in learning more, a great place to begin is to read Salman Rushdie's Booker prize-winning novel *Midnight's Children* (1981), which takes the history of India's conflicting collective identities since Independence as its central theme.

Often nationalists link a national history to a homeland territory, and this represents one of the central tensions in the ideology and a primary point of conflict. The reason is that quite often several nationalisms make claims about the same piece of territory. Consider the troubled and competing views of nationalism in Ireland, or the horrific war that developed in the 1990s after the break-up of Yugoslavia. Across the world similar disputes continue in a whole range of places – Spain's Basque region, the southern Russian region of Chechnya, Tibet and other regions of north-west China. These conflicts are driven by different views of whose nationalism should lay claim to a given patch of territory.

Once you appreciate that nationalism is a modern ideology that has not been around very long, a further important thing to realize is that there are many different nationalisms and these are not simply repeated versions of the kind of nationalism that appeared in Western Europe in the 19th century. Political geographers have argued that there is a tendency to see nationalisms as based on European (or to some extent) American models, but in fact the ideology of nationalism has been transformed and mutated as it has taken hold in other parts of the world. For example, Asian nationalism is argued to have a distinctive form and set of values that are as much to do with pre-existing cultural ideas in many Asian countries as they are to do with Western ideas imported by colonial or imperial powers. Human geographers informed by a postcolonial perspective, and in the aftermath of the cultural turn, are thus interested in how many nationalisms correspond to complex **hybrids** that blend elements of Western nationalism ideology with local qualities, meanings and nuances that never existed in Europe. An example would be the nature of Indian national identity and how it cannot be reduced to a set of values imposed by the British or other European countries, but rather reflect cultural, spiritual and social norms that are distinctive to the many different ethnic and religious communities living across the Indian subcontinent.

For human geographers, one of the key debates over the last couple of decades has been the impact of globalization on nation-states and nationalism. In Chapter 2, we saw that one of the major arguments made about globalization is that nation-states are being undermined or weakened by it. We can now add some more detail to this discussion. The key point here to emphasize is that human geography provides a sophisticated way of understanding why nation-states are not only far from being dead but are unlikely to disappear any time soon. There have never been more nation-states than there are today, and in fact the planet-wide coverage of this political territorial form of state is really only a (very) recent phenomenon. Dozens of new nation-states were born as recently as the 1990s, after the collapse of the Soviet Union. However, that does not mean that political globalization has not changed the capacity of states to govern their territories, or that cultural globalization has not created new dynamics in the ideological basis of nationalisms. It has been argued that nation-states have been 'hollowed out' by globalization, as their powers over the economy moves to local governments, transnational firms and supranational organizations such as the International Monetary Fund, but geographers also point to the consolidation of nation-state powers in other ways (they are still the major holders of power in attempts to govern or regulate climate change or international trade). Similarly, while some point to the rise of global or transnational cultures as phenomena that potentially undermine the strong sense of belonging of populations within nation-states, a geographical view suggests that some of these international flows and cultural dynamics are not that new and that many other forms of everyday practice continue to reproduce the nation-state. Think of what happens when you arrive at an airport, and all the paperwork and systems that states have concerning passports, immigration and national regulations. It may be true that nation-states in the 21st century are being constantly challenged by globalization, but in many other ways they have become more established and permanent aspects of global society than ever before.

THE STATE AND THE ECONOMY

In today's globalized world, states are far from in decline but have evolved a complex set of relationships with other economic actors. In the 1990s political scientists and others argued that nation-states were 'dead' and 'obsolete' (Ohmae 1996), and that they had become irrelevant in relation to a globalizing economy. Economic and political geography have in a range of ways shown how this is not the case, and how states remain crucially important actors in the operation of the global economy. As the economic geographer Peter Dicken argues, in the 21st century states still matter enormously. Their role may have changed from in earlier decades, but states remain at the centre of economic activity not at its margins. Dicken and other geographers identify at least four ways in which this is the case (Dicken 2011).

The first is around the issue of regulation. Economic activity does not exist in a vacuum, and states are important regulators of what goes on. Firms have to obey state laws, and states impose restrictions on what they can and cannot do. Equally, states are central to the regulation of markets themselves. This happens in all kinds of ways, but in our capitalist world, states have the responsibility and power to make sure that markets are 'free' and that individual actors do not have the power to dominate them to their advantage (for example, by trying to become monopolies that dominate an industry or market). Likewise, states impose rules about what kinds of goods and services can be provided, particularly in relation to such issues as health and safety as well as banning certain kinds of goods (for example, addictive drugs). Economic and political geographers, however, are interested in how in today's globalized economy, new kinds of states ('super-states') have joined with nation-states to act as regulators.

A second aspect of the interaction between states and the economy that geographers are interested in concerns the role that states play as 'containers' of markets for goods and services. Economic geographers point to the fact that economic globalization produces a two-way power relationship between TNCs and states, rather than enabling TNCs to be dominant. TNCs need access to the national markets that states are the political institutions in charge of, and in that sense states still have power. It is important to realize, however, that they no longer have the same kind of power as they did in the 1950s when most firms in the wealthier countries made and sold products within national economic territories, and in many countries such as the UK, France or Italy firms were owned by state governments (that is, 'nationalized').

States also act as containers in another way because of the persistence of distinctive cultures associated with individual nation-states which shape the way in which economic production is undertaken in different places around the world. Geographers have made extensive use of a wider social scientific literature that suggests capitalism still comes in a variety of flavours, if you like. This so-called 'varieties of capitalism' literature argues that firms, economic institutions, rules and practices within different national economies mean that economies operate differently in different countries and that this influences how successful (or not) these national economies and firms are at generating wealth.

Finally, states act as both competitors and collaborators in the complex global economy that exists today. They are competitors because they try to maximize their wealth through the best trading position they can achieve internationally. States try to attract investment from firms to build factories, create jobs and thus increase the amount of goods and services produced within their territorial area. However, geographers are also interested in the nature and patterns of collaboration between states in the global economy as they try to maximize the welfare of their populations through relationships that are both political and economic in nature. The major way in which states do this in today's world is through regional trade agreements (RTAs) and forms of regional economic integration. In the case of the former, there are many trade agreements between states in different regions of the globe that essentially offer favourable terms of access to national markets for neighbouring states. The main way this is done is by reducing the amount of taxes (known as tariffs) that have to be paid by favoured trade partners to import goods into each other's national marketplace. In the late 20th century, there was a rapid increase in the number of these kinds of agreements and at least a third of all world trade now takes place within the area covered by an RTA. Regarding the latter form of collaboration, there are at least four kinds of regional economic integration which are politically negotiated between states: free trade areas, customs unions, common markets and economic union. Listed in this order, they represent progressive greater degrees of economic collaboration between states. The aim is to increase the wealth-generating capacity of these regional blocs of states by making industries more efficient through internal competition and economies of scale. There are now many examples ranging from free trade areas such as the North American Free Trade Agreement (NAFTA) and the Association of South East Asian Nations (ASEAN) to economic unions like the European Union (EU).

THE EUROPEAN UNION

The European Union is an economic and political supranational body comprised of 27 member nation-states. The EU had its origins in the European Coal and Steel Community (ECSC) and the European Economic Community (EEC), the latter created by the signing of the Treaty of Rome in 1957. Six countries were founding members: France, West Germany, the Netherlands, Luxembourg, Belgium and Italy. During the 1960s, the early institutions merged but it was not until the 1970s that the community grew. It was enlarged from six to 12 members with the accession of the UK, Denmark and the Irish Republic in 1973, of Greece in 1981 and of Portugal and Spain in 1986. During the 1990s, followed by referenda, the 12 were joined by Austria, Finland and Sweden but with the collapse of communism in Eastern Europe, many East European countries stated their desire to join the EU from the outset and a further ten states (including also Malta and Cyprus) enlarged the EU to 25 member states in 2004, with Romania and Bulgaria also joining in 2007. A political debate is now continuing around the increasingly contentious issue of whether further states should join the EU, most notably Turkey.

During the 1960s and 1970s, the organization evolved from a free trade area into a common market. However, a process of what is known as 'deepening' of integration during the 1980s led first to a customs union with the progressive removal of border controls and then to full economic union, culminating in the introduction of the euro as a single currency in 2000 (although only 17 of the member states have adopted the currency). However since the global economic downturn of 2007, the EU has faced a series of crises in the economies of weaker member states such as Ireland, Spain and Portugal within the eurozone, creating doubt about the continued viability of the single currency. These problems illustrate the challenges that face all forms of regional economic integration in promoting common economic policies where the economic performance of different member states varies significantly. In the second decade of the 21st century, the EU thus faces a major dilemma as to how to sustain economic integration without further political integration.

INTERNATIONAL SECURITY AND TERRORISM

The world in which we live is, of course, neither a peaceful nor conflict-free one. The history of nationalism and nation-states since the 19th century, as with earlier periods in human history, is one full of conflict and wars. The global political map that you can look at in any atlas at the start of the 21st century is thus a product of many centuries of conflict between different groups of people in different territories. Today's national borders are largely the product of previous historical conflict that has produced agreed boundaries between different national communities. Of major significance are the First and Second World Wars, which represented the largest and most geographically extensive conflicts in human history, and led to the establishment of many new nation-states.

Political geographers are therefore interested in the geographical factors that shape the stability and degree of conflict (or lack of it) in the international political system. In the aftermath of the Second World War, a new world political order emerged that sought to maintain international security and prevent future global-scale conflict. The world political map in 1945 was dominated by two major 'superpower' rivals – the United States and the Soviet Union – that each held significant influence over other nation-states. With the end of the Cold War in 1991 (see box in Chapter 2), there emerged what has been described as a 'new world order'. The US remained as a lone superpower but, since most of the communist world embraced free market capitalism the international security of the Cold War era has also been challenged. Political geographers are particularly interested in how these security challenges correspond in large part to a re-scaling of security issues in global society. On the one hand, at the nation-state scale, the stability provided by the two superpowers in the Cold War has been eroded as individual states have challenged international security in advancing their own interests. A growing number of nation-states have developed nuclear weapons (India and Pakistan), and others are seeking to do so (Iran, North Korea), which represents a threat to international security. Furthermore, the capacity of either supranational institutions (the **UN Security Council**) or the remaining superpower (the US) to act as a global policeman has also arguably been reduced with military interventions at preventing 'rogue states' from destabilizing international security meeting with at best limited success (for example, the **Gulf War** of 1990–1). Conversely, the capacity of potential new 21st-century superpowers in the form of China and India to take on a policing role, at least in Asia, is increasing.

Finally, new kinds of threats to international security emerged at the end of the 20th century from above and below the scale of nation-states. This represents another element of human geography's interest in (political) globalization, and the re-scaling of political movements and identities in today's world. At the forefront of this is the emergence of what has been termed 'global terrorism'. As an idea and practice, terrorism is not new, of course. It dates back to the so-called 'Reign of Terror' (1793–4) during the (second) French Revolution and refers to the unlawful use or threatened use of force or violence. This can be either by one person or an organized group, and terrorists target people or property with the intention of intimidating or coercing societies or governments. Terrorism is usually grounded in ideological or political motives and historically has involved activities including assassinations, bombings, random killings and hijackings. It has therefore been a threat to the security of nation-states for several centuries, but it is only in recent decades that it has scaled up to the international level. Global terrorism thus entails attacks by terrorist organizations against many nation-states and the international community, rather than a nationally based one. Foremost in the emergence of global terrorism is the Islamist organization al-Qaeda, which was responsible for the 9/11 attacks on the World Trade Center in New York and the Pentagon in Virginia, as well as various attempted attacks in the Middle East, Europe and Australasia. The rise of this type of globalized terrorism prompted the then US President George W. Bush to declare a new international 'war on terror' in the aftermath of 9/11.

AL-QAEDA: A GLOBAL TERRORIST ORGANIZATION?

Since the late 1980s, al-Qaeda in particular has provoked debate regarding the scale and nature of terrorist activity in the globalized world we live in. The group came to international prominence with the two attacks on the World Trade Center in 1993 and 2001 – the latter being the most devastating terrorist attack in history. It was 9/11 that led many commentators to argue that a new era of global terrorism has begun, since terrorist groups such as the **IRA** had previously tended to focus their activities within one nation and their ideological basis similarly had been related to the politics associated with that area. Al-Qaeda, on the other hand, makes global-scale claims about the nature of its political basis (the intervention of the Western countries in Islamic states around the globe) and has always sought to undertake terrorist activity globally. Importantly, however, there remains debate about to what extent it really represents a 'global organization' as opposed to a regional or national one with membership comprised of cells across the global Islamic diaspora. Groups around the world claim to be part of al-Qaeda but it is not clear that there really is any central hierarchy of command, especially in light of the killing of Osama bin Laden by US security forces in 2011. It may therefore be more a loosely associated set of groups that see advantage in claiming to be part of a coherent global organization when in fact they are acting more or less independently. Human geography thus sees the emergence of al-Qaeda in the wider context of the complicated development of political globalization, the globalization of ideas and cultural values and new forms of global political struggle both above and below the scale of nation-states. They would point, for example, to the way in which representations of al-Qaeda retain echoes of colonial discourses on the irrationality and dangerousness of the Arab and Eastern world (see the discussion in the next section on imaginative geographies).

CULTURE

After discussing how human geographers are concerned with the development of nation-states and their relationship with the economy, it is a natural next step to turn to the issue of culture. Culture has become an increasingly central concept within human geography in the aftermath of the so-called 'cultural turn' in the 1980s and 1990s. Cultural geography has thus become one of the fastest-growing and arguably most dynamic sub-disciplinary strands to the subject over the last couple of decades, and cultural ideas are increasingly permeating many areas of the discipline that previously paid little attention to this dimension to social life.

Culture is a notoriously difficult concept, with academic definitions running into the hundreds. Put simply, culture is a system of shared meanings based around things like language, religion, communities, customs, ethnicity and other identities that are present in all human life. Culture is therefore everywhere, and present in everyday life; for geographers, there is no distinction between the popular (mis)conception of 'culture' as fine art, theatre or opera versus the rest of what meanings people share and how that shapes what they do in everyday life. Culture then exists everywhere at a variety of scales, and is dynamic and constantly changing as people's shared meanings interact and change through time and space. In that sense, culture is unavoidably something that everyone on the planet is involved in rather than a specific or limited 'thing' we do or do not possess (Crang 1998). It is simply 'what humans do' and is thus a kind of process with multiple forms rather than an explanatory variable or a single cause. That does not mean, however, that it does not have 'real' manifestations; whether in the architecture of a city, the products we buy in supermarkets or the films we watch in the cinema, culture is all around us in the material world and people's behaviours and practices. This is the human geographer's view of culture today (as distinct from earlier ideas of culture within human geography discussed in the Introduction).

To understand the different strands of cultural geography, the following section draws out three different major areas: imaginative geographies, the consumption of places, and spaces of consumption.

IMAGINATIVE GEOGRAPHIES

The term 'imaginative geographies' is widely linked to the ideas of the social theorist Edward Said used to describe the ways in which other places, people and landscapes are represented and how these so-called imaginings reflect the desires and preconceived ideas of their inventors (cf. Said 1979) and thus shape action. It is also therefore concerned with the power relations that exist between the inventors of ideas and the subjects of their imagination. Said's work is concerned with the history of colonialism and **Western imperialism** since the 16th century. Said argued that non-Western cultures (and especially those of the so-called '**Orient**') have often been represented by people in the West (in Europe and North America) as being backward, static and inferior. Importantly, he argues that this is tied up with how Europeans and others in the West have historically seen themselves as dynamic, progressive and hence superior. This has led to a set of imaginative geographies that oppose the West against 'the Rest' that, he argues, have played a hugely important role in the nature of global politics and human history in the last two centuries. The West's identity is based on the opposition of a civilized European world that is the absolute opposite of an uncivilized non-European 'other' (see Introduction and Chapter 7). Such arguments echo the wider postmodern debate about identity around the concept of self and other we encountered in the Introduction and will again discuss in Chapter 7.

Said's arguments have been controversial within and beyond human geography because they pose huge challenges to the entire way in which we understand global history and the nature of culture. Important here is the concept of representation, as also discussed in the Introduction. It refers in short to the cultural practices and forms through which people interpret and portray the world around them. Clearly Said's proposed representations are not the only ways Europeans in the last two centuries have imagined non-Europe, but his approach has been enormously important within human geography in the aftermath of the cultural turn. Cultural geographers have now taken the concept of imaginative geographies and applied it much more broadly to a whole range of issues in the world. However, it is important to realize that in essence all geographies are 'imaginative' insofar as they are abstractions that are socially constructed. In that sense, the use of the word 'imaginative' sometimes causes confusion because it is associated in the popular mind with an individual's imagination and even fantasy, implying unreality. Just because geographies are 'imagined' does not mean they do not have real manifestations – the use of the word 'imaginative' is used to denote that these are representations rather than meaning that imaginative geographies are in some way 'unreal'. Driver (2008) makes the point that the study of imaginative geographies is about taking images of spaces and places seriously as things that affect actions and outcomes in the world. Just because the Third Reich's representations of a German state needing more living space in the early 20th century were social constructions does not mean they did not have real effects on the world. Imaginative geographies are therefore 'real' not because they correspond to complete or accurate reproductions of the world, but because they both reflect and sustain how people imagine that world and have real effects.

CONSUMPTION

The issue of consumption has become increasingly important in human geography. If economic geography in an earlier period was focused on the location of production activities, then today human geography has become very much concerned with the geographical nature of how goods and services are consumed after they have been produced. Consumption can be defined as the use of all the products that people create through labour – that can be material goods (whether it is food such as a Big Mac or a device such as an iPod) or it could be a service provided to you (staying in a hotel or having a haircut). For human geographers, the important issue here is that consumption is always 'profoundly contextual' insofar as it is 'embedded in particular spaces, times and social relations' (Crang and Jackson 2001: 2). As a social practice it is linked to people's desires (which are shaped by a range of factors including pleasure, social status and sense of identity), and from a historical perspective it is argued to be linked to the development of modernity since the 17th century and the emergence of an industrial capitalism that enabled a mass consumer society to develop during the 20th century. However, despite its enormous significance, in human geography and some other social sciences little attention was paid to consumption and it was rather simplistically represented as something that simply 'followed' production. Things got made and people then consumed them. It is only in recent decades that human geography (along with other subjects) has shifted to a much more sophisticated view of the production-consumption relationship. It is now appreciated that this relationship is complex because although consumption may follow production, the latter also depends on consumption. Furthermore, human geographers and other social scientists widely agree that consumption does not correspond to an inevitable final moment at the end of a one-directional chain of economic production.

Undoubtedly part of the growing interest human geographers have taken in consumption is the growing importance of it in the world economy. Retail industries have expanded enormously over the last 50 years, employing a growing proportion of the workforce in countries' economies and increasing in power over more traditional 'production' industries such as manufacturing through supplier chains. Think of the classic examples of 'global' clothing brands you will find in any shopping mall or airport whether you are in London, Paris, Los Angeles or Hong Kong. The actual difference between a pair of jeans, trainers or a shirt produced by one firm as opposed to another is very much secondary to how they create an image that entices people to consume the brand. You wear a certain brand because it is 'cool' or 'fashionable' – that is it is perceived to be desirable. People buy Nike over Adidas because they like the lifestyle image it creates. You might buy an Apple computer as opposed to a PC because of its stylishness. And of course a whole industry in itself exists around the creation and maintenance of these consumption images, whether conventional advertising or the endorsement of products by celebrities. Consumption then is a crucial part of our everyday lives and one that has become increasingly central to the world economy and society. While, as an issue, it certainly concerns sociologists and economists who have gone as far as to argue that 'it is consumption, not production, that is the central motor of society' (Corrigan 1997: 1), human geographers have had an enormous amount to say about consumption as an inherently (and increasingly) geographical phenomenon.

THE CONSUMPTION OF SPACE AND PLACE

We need to consider two of the major aspects to this recent geographical interest in consumption. Geographers have argued that consumption often occurs in specific sites and in essence 'makes' place, but equally in today's world it is bound up with the processes of globalization. At the cross over between economic and cultural geography, it is important to consider two strands of the debate around the spatiality of consumption.

First, there is a conceptual analysis within human geography about how the consumption of spaces needs to be understood as a complex interaction between the global scale and local contexts. As previously discussed, a key aspect of globalization is the transmission of ideas, values and practices at the global level. Consumption practices are heavily caught up in this. Consider, for example, how global branded goods are consumed differently in a vast number of spaces across the planet. Some commodities are also consumed in non-material spaces – feature films or computer games that are consumed in virtual spaces. An important aspect to this debate is the degree to which modern consumption is a form of cultural imperialism whereby Western products (particularly American ones) are imposed on the rest of the world. Such products are sometimes argued to promote Western values over others, and thus to produce a loss of 'local' or authentic cultural difference. We have already encountered this kind of argument in Chapter 2, around the idea that cultural globalization is producing homogeneity (cultural sameness) across the world. Consumption activities are seen by many within and beyond geography as a leading mechanism by which cultural globalization occurs. Such an argument is a controversial and contested one, but nevertheless accounts for the growing interest by geographers in the power relationships that underpin consumption. In that sense, human geography has become ever more interested in the broad spatiality and geographies of consumption that many theorists see as being central to the development of global economy and society.

However, closely related to the issues of the global-scale geography and the spatiality of consumption is the issue of how consumption occurs in 'places'. Human geographers have been fond of the idea that consumption 'takes place' to express not only how it occurs as a practice happening in certain places across the planet but also how it is actually recreating and transforming those places. The key argument here is that our imaginative geographies attach attributes to specific places which we consume, and global capitalist consumer culture produces and sells global arrays of cultural and geographical differences. Crang (1998) argues that rather than eradicating cultural difference and producing some kind of 'end to geography', consumption in today's world thus '(re)produces geographies, framing certain local places of consumption as global centres' (Crang 1998: 386). The classic example of this is the emergence over the last century of theme parks such as Disney World. These parks are organized into thematic consumption spaces based around imagined geographical–cultural differences: 'Wild West lands' based on an imagined geography of the American West of the 19th century or 'Adventure lands' based on imagined geographies of European colonialism (think of pirates and treasure islands or jungles and exploration) (Bryman 1995). Another good (if extreme) example is the way in which Las Vegas has developed the theme park idea with casinos along its famous strip themed as mini, stylized, consumable versions of places – Paris, New York, Egypt. More recently, much the same has happened in Macao, the 'Las Vegas of Asia'. Cultural and urban geographers have argued that these theme park 'places' are in fact model examples of a wider process happening beyond their boundaries. A more everyday example is the design of shopping malls as themed consumer spaces that have also begun to look like theme parks, in that instance quite clearly related to the goals of global capitalist firms (see box).

THE SHOPPING MALL: CATHEDRAL OF CONSUMPTION?

Cities across the globe are increasingly dotted with large shopping malls and more are being built every year. Whether you live in the developed or developing world, the mall is becoming a major feature of urban landscapes. Much work in geography has considered these consumption spaces, likening them to 'cathedrals' of global capitalism. Malls are places designed and built specifically for consumption, and evolved from the early shopping arcades that first appeared in big cities such as London, Paris and New York at the end of the

19th century. Malls are designed to maximize the exposure of people to the goods and services being offered for consumption, and are purpose-built material spaces exclusively dedicated to consumption (as opposed to urban shopping streets that also have other functions). Geographers are especially interested in this interaction of material space, architectural design, economic practice and cultural attributes. Malls incorporate specific physical design features aimed at forcing people to pass as many shop-fronts as possible: escalators that make you loop along long stretches of shop frontage, maze-like pathways that do the same or public seating designed to prevent people from lingering. Airport terminals often employ the same tactic as they have become more mall-like, forcing passengers through complex duty-free shopping areas that are hard to escape from (the idea was also first developed by Las Vegas casinos that make it hard for gamblers to find an exit through a maze of tables and gambling machines). The point is that the whole premise of these spaces is to encourage consumption whether that is through thematic decor and images or by maximizing people's exposure to products.

However, there is a further aspect of the consumption of place we need to mention. That is the way in which increasingly *all* places (not just ones specifically designed for the purpose) are being packaged up (or 'commodified') as things to be consumed. One important factor behind this is the growth of tourism as a global industry. Think about guide books such as the *Lonely Planet* series, which now provides near-comprehensive coverage of every country worldwide. As with other guide books, specific places are identified that tourists go to consume visually and experience. The place is consumed by you as a visitor as you gaze at a famous landscape or landmark such as the Eiffel Tower, or experience the 'atmosphere of a place' such as Chinatown in San Francisco. Most books in that series offer the 'top 20 places' to visit, the 'must-see' places to be consumed by you as a tourist. If you go to Australia that is likely to include Sydney Harbour and Ayers Rock (Uluru), or to Thailand the Imperial Palace in Bangkok or the beach at Phuket (the latter being a famous location from the James Bond film *The Man With the Golden Gun*). These material places also quite often become physically altered to meet the expectations of tourists. The point is that places across the planet are becoming things that we know about and expect or desire to consume. This consumption of place as a phenomenon is connected with a range of transformations in today's globalized world: the growth of the retail, tourism and other leisure industries, the deliberate marketing of places as a means of achieving economic growth and regeneration, improved travel and communications industries and of course global cultural flows.

LIVERPOOL, BALTIMORE AND BILBAO: PLACE CONSUMPTION AS A STRATEGY FOR ECONOMIC REGENERATION

Since the 1970s, many industrial cities in the countries of the global North have suffered from deindustrialization (see Chapter 5), and from the 1980s, urban planners and policy makers have made use of what is known as 'culture-led' regeneration. In essence, this is about reversing economic decline by turning industrial cities into attractive places people want to consume. This deliberate planning strategy involves a range of strategies for changing the nature and character of these cities as places in order to promote consumption. It has obviously involved the physical transformation of areas of the urban fabric through a mixture of restoring old buildings for new uses and the construction of new buildings and developments. During the late 1980s, Baltimore in the United States was one of the first cities to develop this approach. Planners in Baltimore focused on regenerating the old industrial and dock waterfront area of the city, which was largely abandoned and derelict, explicitly aiming to create a place within the city that people would come to enjoy retail, leisure and recreation. The image of the place as an attractive one that people would come to consume as a landscape was an important central element of the strategy. People would come to Baltimore's waterfront to live, shop or undertake cultural activities in a landscape that they found aesthetically pleasing.

This approach has now been replicated in many cities across the world as an almost universally accepted way of marketing places to achieve economic regeneration. In the UK, one of the best examples of a similar waterfront place to be consumed is the Albert Dock area of Liverpool. Over the last 30 years or so, this process has involved the restoration of old Victorian docks, turning them into retail and cultural leisure spaces (boutiques, art galleries, hotels, museums) that attract both local residents and tourists to consume a reconfigured place. Museums play on the significant of the city's maritime history and its fame as the birthplace of The Beatles. Art galleries also try to establish the area as a tourist destination. Bilbao in northern Spain is another example where, in this instance, a newly constructed attraction has altered the international image of the city, making it a desirable place to visit and consume. Here the construction of a startling new Guggenheim art gallery on the waterfront forms the focus and has been very successful in changing Bilbao's international image from a decaying industrial town that would have been the last place any tourist visiting Spain would think of going to. Bilbao has been successful in creating an image of a trendy, **avant-garde** place with a cutting-edge art gallery and a 'cool' image. The consumption of this kind of place is thus attractive as a contrast to existing images of tourism in historic Spanish cities such as Seville or Granada. (See also the section on urban regeneration in Chapter 5.)

LANDSCAPE

As discussed in the Introduction, the regionalist approach in geography developed a tradition of recording and representing the features of different regions around the world. Landscape was always recognized as a composite that obviously included material aspects of the Earth's surface (the land) but that human beings also had a central role in creating. During the 20th century, cultural geography developed this idea as its basis, making close linkages between the people living in an area and the form and development of landscapes. Earlier cultural geography essentially saw any given landscape as a gradual outcome of people with a certain culture living in it over long periods of time. Landscape was a kind of record of cultural change, with its form changing incrementally as cultural values change (Crang 1998). This kind of geographical approach to landscape is associated with the work of a group known as the Berkeley school at the University of California between the 1920s and 1950s and its leading figure, Professor Carl

Sauer (1889–1975). Another example of this kind of approach is the work of the 1950s British economic historian W. G. Hoskins (1908–92) and his famous book *The Making of the English Landscape* (1955). Hoskins's account is based on his in-the-field observations of the evidence in the English landscape of past human culture – whether that is the origins of the famous dry stone walls in the Yorkshire Dales or the way in which the feudal field patterns of medieval period remain in southern England.

Since the 1970s, human geography has developed a very different approach to landscape that reflects the radical transformation of ideas from the cultural turn and postmodern or poststructural philosophies that have permeated the subject. Put simply, human geography has shifted away from the idea that landscape can be understood as an observable, external whole (Wylie 2007). Human geography in the 20th century approached landscape as a kind of 'field science' based around geographers standing in the field and observing landscapes to collect 'facts'. Since the late 1970s, geography has increasingly questioned whether neutral observation of a landscape is in fact possible. Instead the subject has sought to conceptualize the *qualities* of landscape – that is 'landscape as a milieu of cultural practices and values' (Wylie 2007: 5). Today, human geography does not conceptualize a landscape as a set of observable 'cultural facts' in any simple or straightforward way.

Within this shift, at least two major strands to this thinking are evident in the subject in recent decades. The first approach to landscape involves an adoption of ideas (and particularly methodologies) from the arts and humanities. The key idea is that landscape can be 'read' in the way you might read a book or a text in order to understand the nature of the societies that were involved in creating those landscapes. By the 1980s, human geographers, had begun to conceptualize landscapes as 'signifying systems' – that is, an array of symbols showing the values through which a society is organized (Crang 1998). Such an approach has been applied by geographers to a whole range of landscapes from the unconventional micro-space of the household through to the landscapes of the national spaces that convey ideas of nationalism already discussed in this chapter. With regard to the latter, we have discussed how imaginative geographies are based around certain representations of landscapes and are often used to convey certain kinds of relationships between people, between people and places and around the meanings of identity. However, there is more to John Constable's early Victorian rural English scenes than just a shared geography of British identity. Paintings such as *The Hay Wain* (1821), which you can see in the National Gallery in London, show people farming in a green and pleasant landscape framed by trees and hedgerows in a certain way. The views are panoramic, from a distance, and give the viewer a sense of being immersed in the landscape.

This style of British painting in the 18th and 19th centuries is known as 'picturesque', generally used these days on postcards. However, in 18th and 19th-century Britain, 'picturesque' had a more specific meaning associated with a new importance placed on images of the countryside in art. What was important was the enjoyment of viewing landscape with this sense of distance, and the way the image conveys a feeling of power and authority over the landscape. Cultural geographers have argued that the development was linked to the fact that the British Empire at home and abroad created a need for middle-class landowners to understand their place in developing land. Being able to represent landscapes in this way became associated with good taste and high social standing in British society. (It is also evident in poetry.) Later representations of British landscapes shift to different symbolic meanings with, in the 20th century, images showing vigorous physical activity in the countryside being seen as an antidote to bland, oppressive urban living (Matless 1995).

The key issue is that reading landscapes can reveal both symbolic systems of meaning and the social relations in societies, and cultural geography has developed this approach far beyond its application to historical paintings. Reading household spaces as landscapes reveals the nature of power relations between men and women in a domestic setting, or in the case of national spaces, the architectural form of buildings or planning of settlements can provide insight into the dominant ideas of nationalism in a nation-state at a given historical moment. Think of the royal parks and palace complexes that have historically been built by different states from in both Western Europe and Asia (China, Thailand, Vietnam). Human geography in the aftermath of the cultural turn has thus applied a symbolic approach to understanding landscape from many angles.

Feminist human geographers have, for example, also read historical landscapes in landscape painting to provide insight into the position of women in a given society in the same way that cultural geographers have brought insight into the power relations inherent in Western imperialism from paintings and other landscapes images.

However, more recently, a second strand of geographical thinking has also become increasingly evident in cultural geography which draws on the different philosophical basis around **phenomenology**, and also on methodological approaches within **cultural anthropology**. This approach argues that landscape needs to be understood as a kind of cultural practice that its inhabitants live through and undertake. Cultural geographers have thus become interested in conceptualizing landscape from the viewpoints of those who live in it and actively make and remake it. The concept of 'dwelling' is used in this strand of geographical thinking to conceptualize how what a landscape 'is' is not necessarily just something that can be 'viewed at a distance' but is also a lived experience that people produce through cultural practices.

Overall, landscape is understood both as a complex representation that reflects certain sets of social meanings and relationships between people and also as a material place that people inhabit and is made and remade as they live in it. While landscape in a popular everyday sense is often associated with some idea of the natural world and countryside (see section on the rural and 'rurality' below), human geographers see it as corresponding to a particular way of thinking about and representing the world around us in ways that have both historical and geographical specificity. Landscapes are cultural things, not intrinsic natural phenomena that exist in the world outside of human meaning. This applies whether we are talking about images of landscapes (paintings, photographs, films) or material landscapes themselves (think of familiar images of the English Lake District, the American prairies or the Australian bush as examples). It is also fair to say that the concept of landscape remains heavily debated within human geography and has a number of competing conceptualizations. As Wylie (2007) argues, at the centre of this in human geography is a tension between the complex and difficult issue of whether landscape is something we look at (an image like a painting) or a material thing we live in (and undertake actions in as well as experience).

THE RURAL AND 'RURALITY'

The ways in which human geographers have thought about ideas of landscape and representation brings us to another important debate: the nature of the 'rural'. Put simply, the 'rural' can be defined as areas 'dominated by extensive land uses such as agriculture or forestry, or by large open spaces of underdeveloped land' (Cloke 2000: 718). It can also include places where there are small settlements that are closely related to the landscape and which 'are perceived as rural' (Cloke 2000: 718). However, this last point is key – that what is rural depends on people's perception. Hopefully it should be clear in light of the preceding discussion of how geographers understand landscape as a social construct that what people understand by the word 'rural' will vary enormously according to where they come from and who they are. It is different in different cultures and in different places. Rurality in Western countries has often been understood as an attractive rural space (think of the European rural areas that people visit on vacation), which contrasts significantly with other places in the world (for example, polluted, dangerous and unpleasant urban spaces). Yet in the global South, rural areas may be remote, isolated, dangerous and without amenities and services. Geographers have therefore approached the issue of how we might define a rural area along two simultaneous lines. On the one hand, an empirical approach can develop ways of measuring how 'rural' places are in a functional way, primarily by assessing the nature of land use and other criteria, such as population density. On the other hand, there is a debate in human geography that is conceptual in relation to the nature of rurality, concerned with those issues that we have been addressing in the section: how rural places are constructed through imagined geographies.

One of the major issues that runs through both approaches is that the meaning of rural is based on an opposite – the idea that some places are 'urban'. In the modern word, the idea of rurality is generally used to refer to areas of land that are not covered by towns and cities. The problem is that in reality there is much ambiguity, depending on which criteria or measure you use, as to whether many places are 'urban' or 'rural'. Many cities in the global South, for example, have extensive hinterlands where land uses that look very urban – houses and factories – are mixed with what we traditionally regard as classic rural land uses – notably agriculture. If you go to India or Thailand, many of the areas around large cities exist in the part-urban, part-rural form. Similarly, densely populated Japan is famous for its complex mixture of urban, industrial and agricultural patchwork of land use. If you take the bullet train from Tokyo to another large Japanese city, the landscape that speeds by is a complex mosaic of these different land uses. This has produced terms such as 'semi-urban' and 'peri-urban' to try and describe this. However, more recently the binary opposition of 'urban versus rural' has been recognized as being problematic, and geographers tend more often now to speak of a 'rural-urban continuum'.

Beyond conceptual and definitional debates, human geography is interested in a range of dimensions and processes related to rural areas. This is a substantial area of work within the subject, and it is impossible to go into all the many debates in depth here. However, at least three areas of human geographical work on rurality need highlighting. The first is an interest in the changing nature of rural spaces in the context of wider processes in today's world. For example, geographers have of course been concerned with rural demography and increasing urbanization especially across the global South. More and more of the world's population live in towns and cities in the 21st century than ever before, and issues from the rate of depopulation in rural areas to the impact on rural economies and lifestyles have been of central concern to geographers. Another important process is that of neoliberalization and in particular the impact of the globalization on rural economies. In many parts of the world, the nature of rural areas is changing rapidly as agricultural industries become globalized and transnational agricultural firms increasingly dominate.

Second, human geographers have also become increasingly interested in the politics of rurality, along with new kinds of politics in rural areas. In the wealthier countries of the global North, for example, the development of many industrial and environmental policies in the EU has been strongly influenced by politics based on rural areas (witness the influence of French farmers as a political movement). This applies equally to the politics of rural areas in the global South with, for example, the role of the Zapatista uprising in the early 1990s in one of the poorest and most rural states of Mexico playing a key role in global resistance to neoliberal globalization as discussed in Chapter 3.

Finally, cultural geographers have also set out to examine the dynamic nature of rural cultures, and particularly how globalization and increasing mobility have produced an intermixing of urban and rural people to a degree to which it not clear whether there is any correspondence between rural territorial space and rural social space. Cultural geographers are also interested in how rural places are being commodified and consumed in much the same way as we have discussed in relation to urban places in this chapter. For example, global tourism has led to new imagined ideas of rurality based on 'natural' experiences that tourists can consume. An example would be adventure tourism in rural Australia, New Zealand or the national parks of the US and Canada, where activities such as white water rafting, bungee jumping, wilderness skiing or exploring forest canopy walkways have commodified a kind of rural spectacle and altered the economies of rural areas in these countries (see Cloke 2005a).

SUMMARY

In this chapter we have considered:

- How human (political) geographers have understood the history of states, the nature of nationalism and the evolution and emergence of a system of modern nation-states since the 19th century;
- The way in which a geographical approach enables a sophisticated understanding of the many complex aspects to the relationship between nation-states and economic activity within their territories in today's globalized world;
- The rise of new challenges and threats to nation-states in the 21st century, in particular the need for 'supranational' trade blocs such as the EU and the threat posed by globalized terrorism;
- How culture has become an increasingly important concept in human geography, and how cultural geographers understand culture as everywhere in everyday life, existing at a variety of scales;
- The significance of imaginative geographies, and how cultural geographers have applied the concept to a range of spatial representations of social and cultural life;

- The nature of consumption as a process and how human geography is concerned to understand how we consume places and spaces;
- How human geography has approached the concept of landscape, particularly debates around reading it as a 'text' that is inscribed with representations and bound up with power relations;
- Human geography's theoretical understandings of what it means for a place to be 'rural', along with the changing nature of rural spaces, and the politics of rurality.

FURTHER READING

Dicken, P. (2011) *Global Shift*. London: Sage.
For the relationship between states and the economy, look at Chapter 6, 'The state really does matter'.

Mansvelt, J. (2005) *Geographies of Consumption*. London: Sage.
This book provides a wide-ranging and thorough discussion of consumption and how it has been conceptualized and analysed in human geography.

Woods, M. (2005) *Rural Geography*. London: Sage.
Gives a good overview of rural geographies using a thematic and conceptual approach.

Wylie, J. (2007) *Landscape*. London: Routledge.
An excellent overview of the theoretical debates around landscape from a human geographical perspective. It also provides a comprehensive account of the issues around the concept of representation.

WEB RESOURCES

If you have not seen it before, have a look at the information website of the European Commission, which has a lot on this supranational organization including policy: http://europa.eu/index_en.htm

The Royal Geographical Society's Rural Geography Research Group has a list of interesting reading and activities/events in this area: www.geog.plymouth.ac.uk/ruralgeography/membersh.htm

5

CITIES, REGIONS AND INDUSTRIES

This chapter considers some of the major debates within economic and urban geography around the nature and significance of cities and regions in today's world, and how this relates to the complex geographies of industries in the global economy.

REGIONS

In everyday terms, the word 'region' has two common uses: either as an area of territory within a nation-state or as a larger area usually comprising several adjacent nation-states on the world map. Human geography makes use of both concepts of the region, and it can be very confusing terminology if the scale of the region being discussed is not made clear. In the 20th century, the concept of 'region' developed as human geography tried to become a 'spatial science' (see Introduction). Regions were reconceived as a certain scale, and as corresponding to systems that linked to larger scales such as the national or global. This represented a major shift away from the 19th-century regional geography that saw regions as areas of territory closely linked to local cultures which in turn shaped the nature of the landscape in that area. The region has thus been an important point of conceptual conflict and argument in human geography with recent work in light of the cultural turn rejecting both the earlier descriptive approach of regional geography and the idea that regions could be conceptualized in the objective terms of a spatial science approach. Debates in the subject about the nature of regions are thus today closely linked to those about place, seeing regions as partial, interlinked spaces that include social, cultural, political and economic phenomena.

It is worth elaborating a little more on the use of the region within economic geography where the (sub-national) region has been one of the major focuses of both theorizing and research. The major reason is that since the **Industrial Revolution**, which began in Western Europe in the 18th century, economic activity has identifiably developed in regions. That is to say, while certain smaller localities such as towns and cities have been and continue to be important in economic development (see section on cities below), the rise and evolution of industries over the last two centuries or so has been most commonly characterized at the regional scale. Whether in the case of the historic emergence of textiles in north-west England in the late 18th century or of the development of mass manufacturing industries in the north-east United States in the mid-20th century, it is the region that human geographers have argued to be the key stage upon which some industries take hold. Within and beyond geography, therefore, the concept of the regional economy is firmly established as one of the major geographical units of analysis in economic activity. In recent decades, this apparently close relationship between particular industries and groups of firms within a region has become one of the major debates within the subject. In the current era of economic globalization, the idea that groups of firms located close together within a geographic region has become one of the major debates within and in several other social science disciplines (notably economics) as well. One of the central ideas is that it is regional rather than national economies – based on these competing clusters of firms in various industries scattered across the planet – that are the major leaders of production and innovation in the global economy. Such an argument makes understanding regions all the more important in a globalized world, but it is also controversial. We will come to examine this debate about the nature of regional economies shortly, but first we need to understand how human geographers understand the major processes that have shaped regional economic development.

INDUSTRIALIZATION

In the historical development of the world economy, industrialization is one of the key processes of transformation that has occurred in modern times. The Industrial Revolution emerged in Western Europe during the 18th and 19th centuries, and spread across the globe (albeit unevenly). While economic historians have of course much to say on this, human geographers point to the fact that industrialization was and continues to be an inherently geographical process. Industrialization has occurred in specific places in a series of sporadic phases or 'Kondratieff cycles' (see box), which have been argued to be associated with a particular set of new kinds of technologies and their associated industries. During the 19th century, industry developed at the level of sub-national regions with certain regions becoming specialized in particular industries. Early economic geography was (almost exclusively) concerned with this spatial agglomeration of industry within regions. The first wave of industrialization occurred in Britain at the end of the 18th and the beginning of the 19th centuries and was based around the cotton, iron-smelting and coal industries. It occurred in the West Midlands, Lancashire, Yorkshire, the north-east of England and southern Scotland. These British regions maintained their success during a second wave from the mid-19th century, but industrialization spread to new areas and beyond Britain to continental Europe: southern Belgium, the **German Ruhr** and parts of northern France. There was also rapid industrialization in the north-east of the US. By 1890, a further geographical expansion of industrialization occurred across central Europe and into new countries: Italy, Austria, the Netherlands, Scandinavia and, in the far east, Japan.

KONDRATIEFF CYCLES

Economic geographers have made extensive use of theories that seek to understand how capitalism is a dynamic economic system characterized by periods of rapid growth and rapid decline. Foremost in this is the idea of the economic cycle named after the Soviet economist who first identified it in the 1920s. Nikolai Kondratieff (1892–1938) argued that phases of growth and contraction had been evident since the beginnings of the modern capitalist world economy in the 18th century. These cycles were of 50–60 years' duration and associated with the development (i.e. the 'invention') of a particular kind of technology. The first Kondratieff cycle involved early forms of mechanization based on water power and, slightly later, steam engines; it began in the 1770s. The focus was on cotton textiles, iron and coal industries. A second wave spanned the period 1840 to 1890, still based on coal, but now around the iron and steel industries, heavy engineering, shipbuilding and the development of the railways. The third Kondratieff wave is argued to be apparent in a phase from the late 1880s to the 1920s, associated with new industries surrounding automobiles, oil, plastics and heavy chemicals. It is at this point that British dominance diminishes as the US and Germany become the leading nations in this wave. A fourth wave then corresponds to the period from the 1930s to the 1970s, centred on the emergence of the 'knowledge economy' industries: information technologies, telecommunications, jet travel and biotechnology. Clearly, however, both the timing and characteristics of Kondratieff waves are debatable. Some argue, for example, that the computer revolution along with the emergence of the internet represents part of a fifth Kondratieff cycle. However, whatever periodization or characterization is used, the broad theory has been widely utilized within geography, and is particularly linked by economic geographers to the geographical unevenness of economies and the rise and fall of urban, regional and national economic spaces.

In the 20th century, the next wave of industrialization was based around different kinds of industries associated with the emergence of mass-produced consumer goods. One of the most significant and discussed industries here is the automobile industry. Companies such as Ford were founded and dominated the economies of certain regions – the north-eastern area of the US from New York, Boston on the coast to Chicago and Milwaukee on the western side of the Great Lakes. However, in what is sometimes considered to be a fifth wave of industrialization based around high technology industries since the Second World War, new regions have come to the fore that are not necessarily those that were industrialized in the mass manufacturing wave, such as the Silicon Valley region of

California on the basis of computers and software. Conversely, just because a region became industrialized around one wave and one particular set of industries does not necessarily mean it will not experience successive industrialization processes.

Historically, therefore, the process of industrialization was extremely spatially concentrated with certain regions experiencing rapid economic growth as industry emerged while others saw very little. In Europe, for example, Spain and Portugal experienced virtually no industrialization in the 19th century, as was also the case with eastern Europe and Russia. Indeed, although industrialization spread beyond its original areas in the Europe and North America, large areas of the planet remain with significant industry for much of the 20th century. It is only since the later 20th and early 21st centuries (as the discussion of economic globalization in Chapter 2 describes) that industrialization has become a planet-wide process with formerly developing countries in the global South experiencing extensive industrialization. Today, for example, similar uneven patterns of regional industrialization are evident across the economies of Asia as China and other economies continue to experience rapid industrial growth. Over the last decade, the pace of industrialization in China in particular has been breathtaking. The Chinese economy grew on average around 8 per cent between 2000 and 2010, and in that last year overtook Japan to become the world's second largest economy. By 2040, if this pace of industrial-led growth continues, China will overtake the US to become the world's largest economy. And this rate of growth is not exclusive to China. India has a similar rate of economic growth and ongoing industrialization, and the Asian economies will account for two thirds of all global output by 2020.

Yet while new regional concentrations of industry bring prosperity, as they have for the last two centuries, in the dynamic capitalist world economy it is also clear that regional economies do not necessarily maintain this success indefinitely. There is a darker side to industrialization in the form of its opposite, mirror process: deindustrialization.

DEINDUSTRIALIZATION

One of the major concerns of human geography since the 1970s has been the geography of deindustrialization across the economies of the global North, and in particular the economic, social, cultural and political impacts of this process on people's lives. The concept refers to the decline of industries and their gradual disappearance from regions and other localities. As with industrialization, it has to happen 'in place' and geographers have been particularly interested in the factors that have shaped the unevenness of deindustrialization between regions.

While some of the earliest industrial regions experienced partial deindustrialization from the late 19th century, it is in the period since the 1960s (in the fifth Kondratieff wave) that deindustrialization has been a prevalent process. By the later 1960s, many of the regional economies in the global North dominated by manufacturing and heavy industries (for example, shipbuilding and heavy engineering) were experiencing industrial decline. The factors behind this were competition from new areas of the world (such as the already growing economies of Asia), overproduction and a fall in demand for the goods being produced. This produced high levels of unemployment, poverty and dereliction of industrial areas (abandoned factories and other facilities). The West Midlands region in Britain, for example, experienced a loss of more than half a million jobs in manufacturing in the 1970s and 1980s (Bryson and Henry 2005). Similarly, that band of the north-eastern United States from New York to Chicago became increasingly known as the 'rustbelt' by the 1980s, as heavier industries and manufacturing experienced significant decline. Places that had experienced the dramatic benefits of earlier industrialization now faced substantial problems: high unemployment, rising crime, urban decay and out-migration.

THE DECLINE OF DETROIT

In the late 1950s, the city of Detroit in Michigan gave birth to a new type of pop music known as 'Motown'. The Motown record label, founded in 1960, propelled acts such as Stevie Wonder, Marvin Gaye and the Jackson 5 to fame and fortune. But in fact the name 'Motown' comes from Detroit's nickname, 'Motor City', because by that period the city had already become known as the home of the US automobile industry.

For most of the 20th century, Detroit was dominated by the car firms that were based there: Ford, General Motors and Chrysler.

However, by the 1980s, increasing competition in car-making from elsewhere in the global economy – along with new robotic manufacturing techniques – led to a crisis for Detroit. Car manufacturing declined, with many former car workers made unemployed, along with many more in related supplier industries. The home of Motown became increasingly known for new and less happy reasons: rising crime, social problems and urban dereliction. People abandoned Detroit and moved elsewhere in the US in search of work.

Despite various attempts by the city and national government to regenerate Detroit, deindustrialization has continued into the 21st century. The automobile firms are still there, although General Motors nearly collapsed in the recession of 2009 and Chrysler is now owned by the Italian firm Fiat. However, they employ far fewer people than they used to. Detroit has also struggled to attract the new industries that are now producing growth in the US economy around informational technologies, biotechnology, software and other service industries. In 2011, Detroit still has one of the highest levels of empty and vacant houses and other real estate that no one wants to buy.

A central problem is that new industries have emerged in different regions (the so-called 'new industrial spaces') in the economies of the global North from the old industrial regions. In the US, newer high technology industries have been concentrated in California and in the region around Boston in the far north-east. In Europe, new industrial spaces have also been in different regions to those experiencing deindustrialization – in Britain as the West Midlands has deindustrialized, the south-east region has seen the growth of new industries. In Germany again, the northern areas of the Ruhr have experienced deindustrialization while the southern **Baden-Württemberg** region has been the area of new industrial growth in recent decades.

Geographers argue that this unevenness in the pattern of industrialization and deindustrialization is not random or accidental. Part of the reason older industrial regions do not experience new industrialization is because of the barriers created by the presence of existing industry. This is captured in the idea of regional 'lock-in' or 'path-dependency'. In short, the problem is that old factories and industrial facilities that are concentrated in a region are not suited to new uses, and the workforce in that region has the skills suited to the old industries rather than new industries. The automobile factories around Detroit and the workers who work in them are not suited to the software or biotechnology industries. The point is that the geographical concentration of industries within certain regions makes those economies inflexible and unable to adapt easily to the needs of new industries (whether in terms of production facilities or the labour force that lives there). To understand the reasons for this in more depth, we now need to consider how economic geographers have theorized the nature of this concentration of firms in an industry within regions, and also one of the key issues that affect whether or not industries succeed within a region: the ability to innovate and compete in the global economy.

AGGLOMERATION AND CLUSTERS

The issue of the spatial concentration (agglomeration) of firms and industries within regions (and particular places in those regions) is one of the biggest debates in the sub-disciplinary area of economic geography. The idea is longstanding, dating back to the work of the late-nineteenth-century economist Alfred Marshall (1842–1924). His argument was that those specialized industrial districts in different regions of Britain created by the industrialization discussed in the last section had a distinctive 'industrial atmosphere' and also benefited from what are known as agglomeration economies: the presence of skilled labour, dedicated infrastructure (for example, transport facilities) and the support of specialist input industries (an example would be the presence of firms in the region supplying specialist components). All of these factors essentially reduce the costs of producing goods and services to firms. However, since the late 1980s, geographers have revisited and developed many of these ideas. Central to this revival of interest in agglomeration is the extension of arguments about its benefits beyond the reduction of costs to other kinds of advantages. Geographers now argue that industry agglomeration is important in terms of all kinds of benefits linked with learning and innovation (Malmberg and Maskell 2002). We will return to this issue shortly, but first we need to think about the most dominant concept of economic agglomeration: the cluster.

While obviously a geographical idea, the most influential theory on business clusters in (economic) geography comes from the work of a Harvard economist, Michael Porter. In several books that have been very influential among policy makers, Porter argues that clusters within regional economies are the major basis for prosperity and competitiveness in today's global economy (Porter 1998). There are essentially three reasons for this, related to geographical concentration: increased **productivity**, greater innovation and higher rates of new firm creation within the cluster. All three make individual firms and the cluster overall more competitive in global markets. Porter's model contends that the clustering effect enhances competitiveness based on four main factors. First, successful clusters tend to serve global markets and thus import innovation into the cluster from demand in the wider global economy. Second, the cluster has close linkages between firms and suppliers or supporting firms that enable complex communication and interaction. Third, competitiveness is enhanced by various inputs within a cluster being concentrated in one region including suitably skilled labour or readily available capital to invest in new ventures. Finally, firms' strategy and rivalry enhance the competitiveness of clusters because proximity encourages new **'spin-off' firms** to emerge in the cluster, as well as rivalry between firms encouraging investment and innovation. The closeness of firms in a cluster to each other (proximity) means that competing firms can monitor each other, and this makes them aware of new developments in their industry very quickly.

The classic example of a successful cluster is that of **Silicon Valley**, already mentioned. Because a large number of computer, software and other information technology firms are located close together in this area, the abilities of the highly skilled workforce who work in these firms are collectively improved and continually updated through the provision of specialist training and education. There are close links with universities, but also the overall innovativeness of individual firms is improved by the collective benefit of workers moving between firms in the region, talking to each other and continually seeing for themselves what competing firms are doing. The 'competitiveness effect' is thus collective, an outcome of all these firms being close together within the region. If they were all scattered across the United States, these collective effects would not be present.

The point about this argument it that it applies increasingly to firms in all industries in the global economy, not just Silicon Valley. Whether that is the film industry in Hollywood or Bangalore in India, the advertising industry in London's West End or the fashion industry in Milan, in the global economy, successful industries are very often located in clusters of firms within specific locations. However, while economic geography has been very much interested in this concept of business clusters, it has also developed a critical debate in relation to Porter's theory and to the way it has influenced government policies around the world. Geographical thinking, for example, has questioned the degree to which firms in many clusters really do benefit from physical proximity in any simple way, also arguing that the nature of clusters varies hugely between different industries as well as between the kinds of wider societies they are embedded within in different places. In that respect, geographers have sought to develop more sophisticated understandings of the factors that give firms advantage through agglomeration. In particular, much geographical work has focused on further developing theories of knowledge and innovation and on the broader significance of what has been termed the 'learning region' (rather than simply firm clusters).

KNOWLEDGE AND INNOVATION

Economic geographers, like other social scientists, argue that we now live in a global economy where knowledge has become one of the most important contributory factors to the production of goods and services. This is what is meant by the concept of the 21st-century global knowledge economy. If for much of the 19th and 20th centuries, firms and regions were economically successful because they had natural resources (coal, oil, minerals) or a pool of cheap labour, then in recent decades what has become more important is the level of knowledge and how that knowledge is applied and built upon. This increasing importance of knowledge is sometimes also termed the 'informationalization' of the global economy (or the global informational economy).

Geographers have therefore joined economists and management theorists in putting much emphasis on the relationship of knowledge to economic activity. They have become increasingly interested in this respect in different forms of knowledge. There is an important distinction to be made between *codified knowledge*, which is formal and systematic (the kind of knowledge you find in a manual or a textbook) and *tacit knowledge*, which is based around direct experience and cannot be easily expressed in texts or documents. Tacit knowledge is therefore more practical and represents the 'know-how' involved in doing something. You only have to think about the difference between two people trying to operate an electronic gadget such as a DVD player or games console, one by reading the instruction manual and the other by having experience of the gadget from repeated use. Tacit knowledge or knowing how to operate something makes all the difference.

The role of different forms of knowledge is crucial in relation to a key process in economic activity – innovation. In the context of the economy, innovation is the creation of new goods and services, and it includes the modification of existing ones. Firms in today's world are constantly trying to innovate in order to remain competitive and/or increase their profits. As a process, it is evident everywhere. Think of the latest mobile phone or piece of computer technology. Firms such as Nokia or Dell in these industries continually alter and amend models, incorporating new features and improvements in order to persuade people to buy their products. Innovation also comes in the more obvious form of *product innovation* (the latest Apple notebook) and *process innovation*, where firms develop new ways of making a product or delivering a service. An example of the latter is the range of innovations the internet has produced through e-commerce. Consider how newer, budget airlines (such as JetBlue in the US or EasyJet in Europe) cut the cost of their fares to the minimum by using website-based ticket sales, new staff working patterns and more sophisticated booking systems. Such an innovation means they have to employ fewer staff in airports and in their head offices, which enables these companies to charge passengers less.

Economic geographers and other social scientists tended to understand innovation as a linear process in much of the economy until the 1990s. Such innovation happened in large companies that had formal research and development (R&D) departments, which employed skills engineers or scientists to develop new products in a separate environment from the main divisions of the company that produced its existing goods and services. However, this view has changed over the last couple of decades with the realization that a lot of innovation is in fact interactive in that new products and ideas emerge from the constant cooperation, collaboration and exchange of ideas between customers, suppliers, research organizations (universities) and a whole range of employees in different divisions within companies. All this is important to geographical thinking because of the role of different kinds of spaces and of spatial agglomeration in the processes of innovation. Much economic geography has argued that innovation in firms is facilitated by the proximity that agglomeration brings, and is embedded in a wider range of factors that are specific to places (see box). Firms located in one region are more innovative because of the collective enhancement of learning resulting from them being close together.

EMBEDDEDNESS

This concept essentially refers to the way in which economic actors and activity are caught up in a range of factors that in conventional economic analysis are regarded as 'non-economic'. Economic activity is thus embedded because, so the argument runs, it does not exist in isolation from a whole range of influences in wider society. This includes social and cultural values, laws and regulations, individual and collective behaviours as well as political circumstances. The concept has come into economic geography via a broader interest in a number of social science disciplines (for example, the area of **economic sociology** within sociology or **socioeconomics** within economics). An important basis for this is the mid-20th century thinking of Karl Polanyi (1886–1964), an interdisciplinary scholar who is, broadly speaking, an economic historian. His book *The Great Transformation* (1944) provided an historical analysis of how economies are shaped by culture and the nature of the societies in which they exist. Such a position is highly critical of the dominant, mainstream arguments of neoclassical economics that treat the economy and economic activity as phenomena that can be abstracted from their socio-cultural context.

Geographical work since the 1990s has become increasingly interested in the ways in which industries, clusters and even individual firms are embedded in the socio-cultural context of specific places and spaces. Geographers have become interested in different types of embeddedness: territorial, social, institutional and so on. All of these ideas aim to capture how firms and other economic actors are caught up in non-economic influences. However, what has made the concept of embeddedness especially useful is the ongoing globalization of economic activity, with geographers seeking to understand how transnational firms, global production networks and global commodity chains are all embedded in increasingly complex ways in multiple places (see Chapter 2). Dicken (2011) captures both the importance and the complexity of this when he states that TNCs are produced 'through an intricate process of embedding, in which the cognitive, cultural, social, political and economic characteristics of the home country continue to play a dominant part' but where they also 'inevitably take on some characteristics of their host environment' (Dicken 2011: 122). Equally, debates regarding the nature of clusters and of learning regions show how strongly embedded in the many characteristics of certain places the process of innovation and the competitiveness of firms are. Economic activity in today's world is not only embedded but is so in increasingly complicated ways in multiple places and contexts simultaneously.

THE LEARNING REGION

Economic geographers have used these various theories of knowledge and innovation to develop a broader approach to the debate about clusters and regional agglomeration around the idea of 'the learning region'. At least five arguments are made in relation to this concept. The first is that globalization does not make all places equally attractive for economic activity, but rather leads to new forms of agglomeration based around where knowledge can be created (Storper 1997). With the rise of new and highly effective global information and communications technologies, codified knowledge has become increasingly available everywhere (for example, you can download your games console manual from the internet anywhere). However, and second, the opposite is true of tacit knowledge, which is 'sticky' and remains very firmly rooted in specific places. The reason is that tacit knowledge is something possessed by individuals that can only really be made use of face to face when employees work together in the same place. You cannot very easily transmit it elsewhere. What this means for specific places is that those that can create and maintain tacit knowledge in certain kinds of economic activities will have an advantage over those that cannot (Maskell et al. 1998).

A third issue is that of the informal rather than formal social practices that are involved in economic activity. Regional economies benefit from all kinds of informal relationships and linkages between individuals that tie firms together. The economic geographer Michael Storper called these 'untraded interdependencies' (Storper 1995) that are hard to measure and that loosely correspond to the collection of skills, attitudes, habits and shared understandings which come about in an area where there is specialized production. Such interdependencies are generated by, for example, groups of people working in an industry meeting in bars and clubs, playing golf together or socializing in chambers of commerce and trade associations.

Storper also argues that learning regions also have another similar and hard-to-measure characteristic – what he and Anthony Venables have called 'local buzz' (Storper and Venables 2004). This is a kind of updated version of Marshall's 'industrial atmosphere'. It refers to an ill-defined vibrancy and excitement in everyday life within an industry cluster that is a consequence of many different activities and events occurring in one place that create interesting and potentially useful information and knowledge for economic actors. It is very much dependent on face-to-face interaction and the co-presence of both firms and people within an industry in the same place (Bathelt et al. 2004). Local buzz is most easily imagined in an urban setting. An example that geographers have studied is the advertising, media and computer animation industry in the Soho area of central London (Grabher 2001). The cluster exists in the cramped and trendy narrow streets of this district, which is filled with bars, restaurants, theatres and clubs as well as companies. Firms benefit from just being in that place as employees meet each other, chat, gossip and interact in all kinds of ways from eating out or drinking in bars informally to attending industry events (the launch of a new advertising campaign or a feature film that a firm in the cluster has worked on).

Finally, geographical work on learning regions has emphasized the significance of trusting relationships between firms, which is crucial if they are to collaborate and learn collectively. The idea here is that greater closeness between firms (physical proximity) means they are more likely to trust each other than trust distant firms with whom they have only periodic contact. Economic geographers have also begun to theorize these kinds of trust at the level of individuals. For example, in many industries, such as finance or legal services, trust between specific senior managers (or firm partners in the case of legal services) has been found to be crucial (Faulconbridge 2008).

INDUSTRIAL DEVELOPMENT

Geographical analysis of economies today goes beyond scale-based ideas of national, regional or local economies and also has a considerable interest in the nature of economic activities: namely the development of specific kinds of activities (industries) and how this is also part of the explanation for the geographies of economies we see in today's world. We therefore now need to consider some of the major ways in which geographers have contributed to an understanding of different industries and how a geographical viewpoint sheds light on the roles of different types of industries in the global economy.

MANUFACTURING

It is one of the commonest misconceptions in some wealthier countries in Europe and North America today that, overall manufacturing industries have declined. While it is true that these countries may have experienced manufacturing industry deindustrialization, in terms of the global economy, manufacturing output has increased almost every year since the end of the Second World War. What has changed dramatically at the global scale, however, is *where* manufacturing goods are made, *what* is made, and *how* they are made. Taking the issue of where manufacturing industry is located first, the major trend has been what Peter Dicken calls his 'global shift'. Up until the 1970s, the vast majority of the world manufacturing industry was located in the global North, but increasingly manufacturing moved to economies in Asia and Latin America. By the early 1980s, geographers were interested in what they saw as a new international division of labour whereby manufacturing work was increasingly being done by workers in the global South and other service-based jobs by those in the global North. Since then, this geographical reconfiguration of manufacturing has become ever more complicated as TNCs organize production through global production networks. However, it needs to be emphasized here that, despite this shift towards GPNs, for manufacturing production the trend for this to be located outside the wealthier countries of the global North has continued. Led by Japan, but now increasingly China, Asian countries produce more and more of the world's manufacturing output. Lower labour costs are one of the major factors that have led to this relocation over recent decades, although it is increasingly also about the fact that countries such as China have growing markets for manufactured goods. Geographers have sought to understand the trends in this shift of manufacturing production from the economies of the global North, and explain why some manufacturing industries have continued to succeed in countries like France, Germany and the United States, while others have all but disappeared. For example, Germany still has substantial automobile and machine engineering industries, with companies such as BMW and Bosch thriving. The British and American automobile industries, by comparison, have suffered continual decline both in terms of the number of people employed and the success of firms from those countries. There are now wholly owned British automobile manufacturing firms and the two largest US companies – Ford and General Motors – came very close to bankruptcy in the global economic downturn of 2007–9. Automobile production, however, is still strong and growing in many Asian and Latin American economies. Part of German automobile firms' success, for example, is accounted for by their success in moving into China. Volkswagen sold more than a million cars there in 2010.

Second is the issue of what is made. Over the last 50 years, the nature of manufactured goods has changed radically. Broadly, this has two aspects: a huge increase in the volume of manufactured products made, and a dramatic increase in the number of different types of product. Compared to the 1950s or 1960s, the 21st-century global economy manufactures much more than previous periods. This is part of the reason many manufactured goods have become more readily available and cheaper in 'real terms' (that is, they cost a smaller proportion of people's total income than in the past).

Furthermore, there are many more different products. This includes a huge number of manufactured goods that are simply *new* and related to technological innovations of the fourth or fifth Kondratieff. A good example would be the electronics industries, which now produce numerous electronic consumer goods that did not even exist 30, let alone 60, years ago. Think of CD and DVD players, personal computers, laptops, printers, digital scanners or mobile phone technology. Equally, however, what has also changed is the huge increase in the various types of manufactured goods: you can now buy hundreds (if not thousands) of models of many electronic goods. Consumers have a vast choice compared to previous decades.

This leads us to a third issue in relation to manufacturing: the dramatic changes in how manufactured goods are made. A large body of work in economic geography has been concerned with the shift to what is argued to be a new kind of more flexible manufacturing economy that has enabled both an increase in the volume of production and all that diversity in manufactured products. Geographers in the 1980s and 1990s focused much attention on the idea that Fordist manufacturing was increasingly being replaced by new ways of making goods that were loosely labelled 'post-Fordism'. Economies and societies of the global North in the mid-twentieth century were characterized by a period known as Fordism (named after Henry Ford who founded the Ford motor company), which involved the mass consumption of manufactured goods (everything from automobiles to vacuum cleaners, television sets and refrigerators). This began in the United States in the 1920s and 1930s, but after the Second World War it spread to Europe and then to Australia, Japan and other economies. The peak of this Fordist period is argued to be the 1950s and 1960s, but by the 1970s companies sought to keep up their profits by beginning to shift to cheaper production locations in southern Europe, Latin America and part of Asia. Importantly, however, manufacturing firms also learnt new methods to enable them to be more flexible with the goods they produced. By the 1980s, people were used to manufactured goods coming in many different forms – different colours, models and versions of a product. This required firms to develop many kinds of flexibility in how they made goods. The classic but probably over-used example is automobile production, where firms began to make use of robotic assembly lines that could produce small batches of automobiles with different features. If you buy a BMW's Mini today, you can specify in the showroom not only its colour but all kinds of features – the kinds of seats, alloy wheels, a sunroof, the type of audio equipment it has. All of this is possible because of flexible production methods that allow firms to tailor complex manufactured goods such as automobiles to specific customer preferences.

SERVICES AND THE KNOWLEDGE ECONOMY

In the last few decades, economic geographers have become increasingly interested in the service industries. Conceptually, however, the definition of what corresponds to the 'service sector' is something of a problem. Services used to be regarded as fairly unimportant by human geographers and other social scientists, a kind of small additional set of economic activities that were marginal to the main industries in the primary (oil, mineral extraction) and secondary (manufacturing) sectors. In the second half of the 20th century, this changed dramatically in at least three major ways.

The first was the massive growth in the size of the service sector in most of the world's economies. In the latter half of the 20th century, the overall contribution of industries that are described as services to GDP increased dramatically. For example, taking economies in the global North such as the US and UK, the service sector moved from accounting for 42 per cent of GDP in 1970 to 73 per cent by the year 2010. A major factor has been the development of the global knowledge economy discussed above since many new service industries are 'knowledge-intensive' and reflect the growing need for knowledge in economic activity. However, it is important to realize that the growth in services is neither restricted to the economies of the global North nor just related to the emergence of an increasingly knowledge-oriented economy. Some new services are also a reflection of the growing prosperity in the global economy and the fact that consumers can afford to pay for services they could not previously. For example, think of the growth in domestic cleaning businesses or other home-related services like childcare or even dog-walking!

Another factor is the way large companies have moved away from providing many services 'in-house'. Car manufacturing firms no longer have their own catering or advertising divisions as they may have done in the 1960s or 1970s. Now they buy in these services from other specialist companies. The dramatic growth in services is therefore multi-dimensional and has an uneven geographical impact. Economic geography has therefore been especially interested in the distinct differences between areas where service industries have developed and flourished and those where they have been less evident. A key example of this is the argument that certain kinds of service industries have been central to the success and development of global cities (as discussed in Chapter 4), but geographers are also concerned about how a poorly developed service sector is associated with regional economies that have struggled to create wealth and jobs in recent decades.

A second trend is the enormous diversification of activities that come under the label of a 'service'. There are, in short, lots of new kinds of service industry compared to even 30 years ago. In the 21st-century global economy, the types of industries that we can call services have multiplied many times. In the pre-Second World War era, whole sectors – such as IT services, business consultants and media services – simply did not exist. Moreover, there is considerable debate among economic geographers and other social scientists as to whether many of these industries are 'pure' services, or whether they combine elements that are better defined as something else, such as creative industries (see the next section on the 'new economy').

One important aspect of the diversification of services, however, is the useful distinction between 'producer' (or business) services that are provided by firms to *other firms* and 'consumer' (or retail) services that are offered to individuals. Regarding the former, industries such as accountancy, advertising or management consultancy fall into this category (you would not buy the services of these firms personally), whereas consumer services include a whole range of industries from leisure firms owning gyms and cinemas to classic services such as hairdressers. Some service industries have both producer and consumer dimensions to them – banking, law and insurance are classic examples of these. Geographers and other social scientists have come up with at least seven groupings for various service industries that now exist: finance and real estate, business, transport and communication, wholesale and retail trades, entertainment and leisure, education and health and not-for-profit (charities, museums and galleries). Just the length of this list and the number of types of firms and organizations it covers should give you some idea of how huge the service sector is nowadays.

Finally, the third trend around the service sector of the global economy is the growing importance of certain types of services to the operation of all industries in the global economy. This brings us back to the issue of the knowledge economy and specifically the increasing importance of producer services. Put simply, the growing importance of knowledge as an input into everything that is produced in the global economy means that specialist producer services have become more important. Whether or not the aircraft company Airbus can develop and sell a new model of plane is reliant on the inputs of many specialist firms offering services around a whole range of areas including engineering, design, specialist recruitment, software and marketing. Producer services are thus increasingly involved in every other industry, and that means they can also generate a lot of wealth in the regions and places where they are concentrated. They have also been central to processes of economic globalization as producer services have in effect helped transnational firms to become transnational, and the global economy to become more integrated (see box). We will return to this issue shortly in consider the nature of global city networks.

THE GLOBALIZATION OF BUSINESS SERVICES

As recently as the 1970s, a number of the most important business services – investment banking, accountancy, legal services and management consultancy – were very much based within national economies. In the UK and Germany, for example, banking was concentrated in London and Frankfurt respectively, and in the US, New York, particularly the area around Wall Street, dominated finance. Similarly, law firms in these cities providing services to other companies (corporate law) focused on their national markets and rarely served client firms outside of their home economies (even if they were dealing with investments or business deals these firms were involved with elsewhere in the world). However, since the late 1970s, all of these business services industries have changed dramatically and their activities have become globalized.

Why has this happened? There are at least three major reasons. First, the activity of these firms has been deregulated since the 1970s (especially in finance), allowing them to do business overseas much more easily. Second, and following on, is the logic of firms seeking to grow and increase profits by expanding their operations into new markets. This means offering their services to firms in countries where they have not previously done so. UK law firms, for example, have tried to expand into North America and Asia to offer their legal services to firms. Finally, the globalization of these industries is tied in with the wider general trends of economic globalization. Business service firms help other firms in all sectors of the global economy become more global themselves – for example, investment banks provide the finance for European firms to invest in Asia, while management consultancy firms give advice on how to set up operations and law firms draw up contracts to make the deals happen.

THE 'NEW ECONOMY' AND CREATIVE INDUSTRIES

One particular part of today's global knowledge economy that has been of great interest to human geographers is those industries that are associated with what is termed the 'new economy' and that are in some way 'creative'. One of the reasons that geographers have been especially interested in these industries is partly because, in order to fully understand their development, there is a need to explore different elements of the subject. While the above account of the appearance of new service industries covers many aspects of this 'new economy' of recent decades, not all of the industrial activity associated with this idea fits neatly into the category of 'services'. Many new knowledge-based industries appear to make products rather than provide a service (or at least do both), even if these are not material goods. These kinds of 'creative industries' range from more traditional industries that have evolved – such as advertising and marketing, music and the visual and literary arts – to more truly new activities that have only really come into existence in recent years, such as computer games, film and television, media and web design. They have been of significant interest to economists, business theorists and planners because they are seen as key drivers of economic growth (especially within urban economies), but geographers have also done a lot of work on these sectors. They have, for example, examined what makes successful computer games industries or examined clusters of designer firms in the global fashion industry.

The reason for this may be the argument that a geographical approach is especially useful in understanding the many factors that lead to the development of these industries. Creative industries are impossible to understand without an appreciation of cultural transformations in global society. In that sense, human geographers arguably have the edge in conceptualizing the development of these new economic activities and creative industries over accounts in subjects such as economics. One particular debate in this respect shows this very well: the idea that the people who work in creative industries are the key to economic growth in city-regions. Geographers have been very concerned here with the work of the US policy commentator Richard Florida, who argues that, in the global economy, city-regions succeed if they can attract the skilled workers who are employed in creative industries (Florida 2002). This new 'creative class' includes not only artists and musicians but all kinds of jobs in fashion, media, marketing and so on. One of the arguments is that these creative industries cluster in attractive city environments that people in this creative class work in, which also links to ideas about clusters and 'local buzz' discussed earlier. Human geography is in a particularly strong position to understand the nature and significance of creative industries since success or failure of this kind of economic activity is seen as being bound up with cultural and place-related issues that are normally outside the concern of economists or business theorists. Geographers have investigated the extent to which the characteristics of certain places influence the concentration of creative workers and industries, and examined how policymakers might seek to attract these kinds of people. These issues are also related to a geographical approach to understanding the changing role of cities in today's world, which we will explore shortly, and are especially good examples of how the boundaries between 'economic', 'social' and 'cultural' geography are very much blurred.

AGRICULTURE AND FOOD

A fourth and final group of industries that geographical work has been concerned with provides a contrasting example of the way in which the interdisciplinary nature of human geography is helpful in understanding the complex changes in today's global economy. In the case of agriculture and food production, dramatic changes to the actors responsible for producing food together with the way these industries are organized has brought huge changes and challenges to the landscapes and environments that people around the world live in. Equally, as we mentioned when considering consumption in the previous chapter, geographical work sees the issue of food production as very closely tied to questions of consumption. As Brian Ilbery and Damian Maye (2008) put it neatly, 'food is a geographical topic'. We could argue that it is becoming ever more so as globalization processes further transform the way in which the agricultural industry produces foods, and what foods people around the globe are able and wish to eat. There is therefore a large body of work on food within human geography, but three major aspects of geographical work on agriculture and food are worth highlighting.

First, in terms of agriculture and food production as an industry, a geographical approach is very much interested in where and how food is produced and the complex system by which it is transported and sold to people to consume (through retail). Economic geographers are thus very concerned to map the increasingly complex *global food chains* or networks of food production and distribution that involve transnational agricultural firms and food retailers (supermarkets). The concept of the global food chain traces the multiple connections to different places of production for food and agricultural commodities. An example would be to think how the food you buy in fast food restaurants such as McDonald's or Kentucky Fried Chicken has been produced in different places, and how it is stored and distributed through networks of outlets in an economy until you consume it. Another example is, of course, the global food chains increasingly controlled by the transnational food retailers discussed in Chapter 2. One of the major developments geographers have been concerned with is the push to make food chains more sustainable. A burger restaurant in Europe could be selling burgers made from beef flown from Argentina or Brazil in South America, along with fresh produce such as lettuce or tomatoes that have also been air-freighted from other tropical countries. An understanding of the geography of the food chain therefore allows an assessment of the impact on the environment of certain ways of producing and distributing food (in terms of carbon emissions into the atmosphere in this example).

Another second aspect of geographical interest in agriculture and food is also the social impact of changing food production. With the increasing power and dominance of TNCs in food retailing and agriculture in countries of both the global North and global South in the 21st century, geographers have been concerned with the dramatic changes of livelihood that occur as traditional rural ways of living disappear. Small farmers are being replaced by industrial agriculture across the world, and transnational firms also increasingly dominate through their use of specialist seeds. Furthermore animal species have been developed that have led to an increase in the amount of food that can be produced in a given area of land. At the other end of the food chain, the transnational food retailers such as Walmart, Tesco and Carrefour that we discussed in Chapter 2 have increasing power in the marketplace for foodstuffs that often also further drives the shift to large-scale agriculture. Geographers have been interested in the responses to these changes including, for example, attempts to protect small food producers in the global South through cooperation and **alternative food networks**. Examples of this would include global initiatives to develop fair trade in places where for a long time poor farmers in the global South producing commodities like tea, coffee, sugar and cotton have received very low prices.

Finally, as mentioned already, the changing nature of the consumption of food has been of great concern in cultural geography. The changing nature of food consumption is in part the result of many factors over the last 50 or 60 years, including changing methods of food production and distribution, as well as social and cultural transformation in many regions of the globe. Cultural globalization has exposed many people to new kinds of food, and economic globalization has enabled foods to be distributed to places where they were never previously available. These processes mean, for example, that a wide variety of cuisines are increasingly available everywhere on the planet. You can now as easily eat American, Italian or French food in Tokyo or Beijing as you can find Japanese or Chinese food in Europe or the Americas. While this globalization of food has been going on for centuries, in today's world it has become much more highly developed. Cultural geographers have naturally been very concerned with understanding how representations of places and cultures are bound up with different cuisines and foodstuffs, and how these are changing and developing in the globalized world we now live in.

CITIES

The majority of people in the world today live in cities or places that would be described as 'urban'. In the richer countries of the global North, it is estimated that more than 70 per cent of people live in urban areas and the developing countries in the global South are rapidly catching up with this figure: the comparable figure is already 60 per cent. Human geographers have long been interested in cities, so much so that – as discussed in the Introduction – there is a whole sub-discipline of 'urban geography' within the subject. Yet defining a city is itself tricky. Cities come in all shapes and sizes, and have very different make-ups, in terms of who lives in them. Geographers broadly have made use of a series of criteria for defining an area as urban based on the size and density of the population living in a particular place, how permanent the settlement is and how diverse it is in terms of the types of people living there (Cochrane 2008). The problem remains, however, that the nature of cities in today's world remains very diverse. Cities in the global South, for example, such as Mumbai, Mexico City or Lagos have very different social structures, physical forms and urban politics from those of many cities in Europe or North America. Generalizing about cities is therefore difficult despite the fact that in the second decade of the 21st century, urban geographies are becoming ever more significant not only because more and more of the world's population live in urban places, but also because cities are increasingly the key places where many issues facing the world today come together. Whether because of the fact that 'global cities' are increasingly the main places where global economic activity is organized (Sassen 2001), or because of the need to develop sustainable ways of city living in a low-carbon future (Betsill and Bulkeley 2005), urban spaces are now the key places.

URBANIZATION AND URBAN FORM

The concept of urbanization refers simply to the way in which cities have grown as more and more people have moved to live in them. Throughout most of the last couple of centuries, urbanization has continued at a steady pace. Cities in the year 2000 were more numerous, had bigger populations and covered more land than they did in 1900. The same is true of the previous 100 years if you look back to 1800. This process, however, has not always been either uniform or consistent for all cities in all parts of the world. In the wealthier regions of the global North, large cities in the late 20th century did also experience decline and an opposite process of counterurbanization. However, for the most part, it is urbanization processes that have been dominant.

Both urbanization and counterurbanization are of course inherently geographical phenomena insofar as they generally involve the movement of people to live in cities (see also the section on migration in the next chapter), and the physical growth of cities in terms of land area. Of equal interest to geographers, however, is the wider question of how cities expand into new territorial space, as well as how they change over time. Geographers have therefore a longstanding interest in urban form – that is, the physical structure of how cities are laid out, where buildings are located, what kinds of buildings are in particular areas of cities and what factors have shaped this. In the earlier 20th century, urban geographers were involved in the development of models of how cities had developed in order to understand how different areas of a city have different uses. One of the classic models in urban geography was proposed by the American sociologist Ernest Burgess (1886–1966) in 1925, based on his work on the city of Chicago at the time. Burgess famously argued that land use in the city was organized around concentric zones of usage, with the central business district in the centre and various old or new industrial or residential areas further out (see Figure 5.1).

While historically specific and simplistic models such as this formed the basis for many attempts by urban geographers to understand the form of cities, the sub-discipline remains closely concerned with how different areas of cities gain or lose certain land-use characteristics. In the last 60 years, major debates in this respect have been concerned with how deindustrialization, **flexibilization** (Fordism to post-Fordism), informationalization and globalization have affected urban form. These are enormous debates in and of themselves, but we can briefly identify at least two

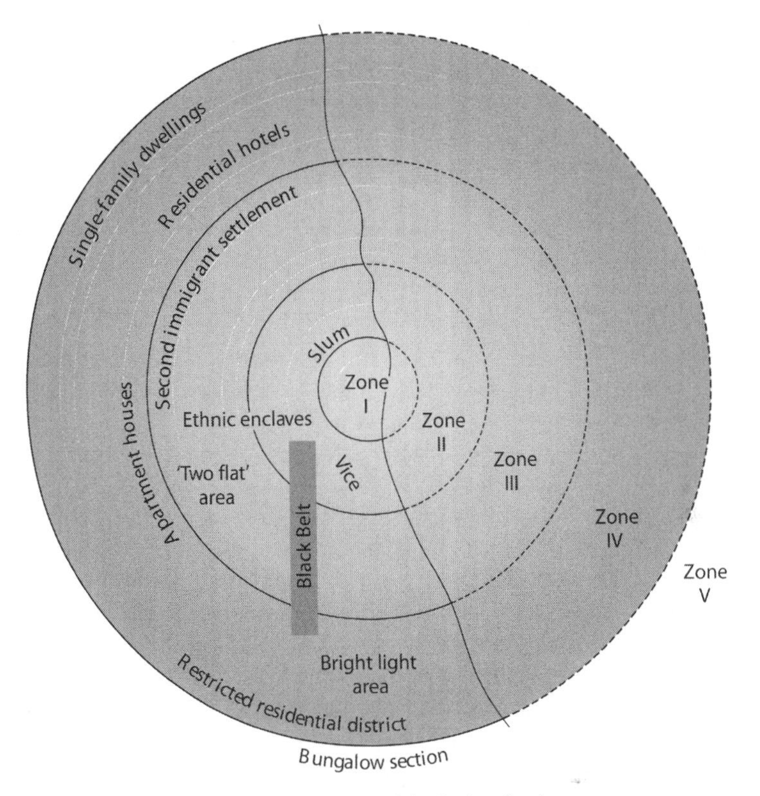

Figure 5.1 Burgess's concentric zone model of urban land use

overlapping features of changing urban form that have concerned geographers.

In relation to economic change, there has been a shift away from the urban form associated with the heyday of Fordist manufacturing in the 1950s and 1960s. Many cities in the global North have experienced an ongoing suburbanization that entails their sprawling over large areas of land and leads to the dilution of clear land use in different parts of cities. Suburbs today in many cities in America, Australia and Asia contain many land uses mixed up together – residential, retail, industrial – reducing the significance of the central area. This mixing up of city form is sometimes referred to as the emergence of 'edge cities' because these suburbs contain everything you would normally expect to find in a city, albeit in a very spread-out form (Soja 2000).

A second set of changes to urban form that has concerned geographers relates to the globalization of cities and their increasingly interlinkage into global city networks. Changing urban form in many cities is a consequence of the relationship a city or area of a city has (or does not have) with the global economy. The construction of new central business district areas (an example would be the Canary Wharf area in London) or the **gentrification** of old poor housing neighbourhoods over recent decades (this applies to areas of both London and New York) is linked to the office space needs of transnational firms and to the higher incomes of specialized service industry employees respectively (Hamnett 2003). Equally, urban geographers have also pointed to how globalization is creating areas of cities in the global North that look much like districts that used to be associated with cities in the global South. Wandering along the street markets of east London or the Chinatown districts of many American cities, it is increasingly hard to distinguish the form and character of these areas from the types of urban spaces that may be encountered in Asian cities such as Hong Kong or Bangkok. To understand such changes in urban form, we need to consider the forces behind such change. This brings us to the concept of urban systems.

URBAN SYSTEMS AND GLOBAL CITY NETWORKS

Historically, the origins of trying to understand cities as part of urban systems relates to the way in which urban geographers sought to classify different types of cities within different countries and nation-states. Such an approach established that larger cities played a more important role than smaller ones within regions and countries. Within human geography, this theory of 'central places' was first developed in the 1930s by the German geographer Walter Christaller mentioned in the Introduction (see Figure 5.2).

Central place theory is based on assessing the functions fulfilled by different urban settlements that were spread across a piece of territory, and argued that larger towns and cities provided more important, rarer services to the surrounding area. In the later 20th century, and especially after the Second World War, urban

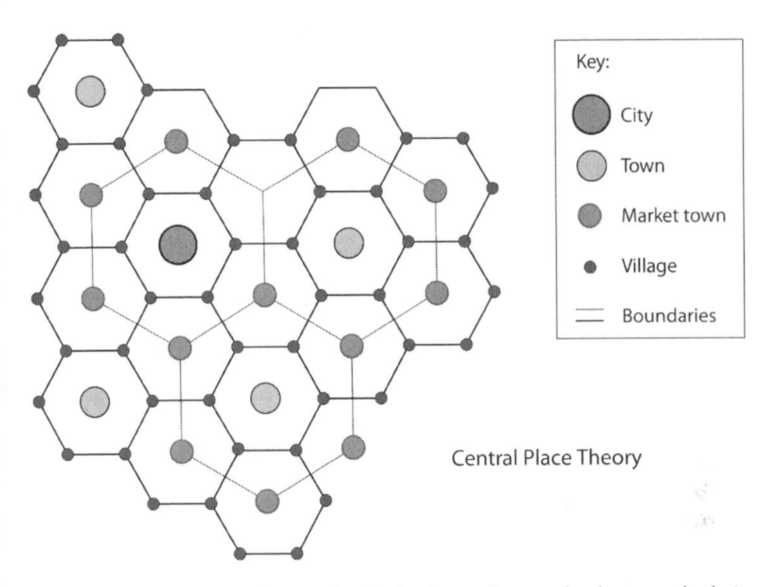

Figure 5.2 The size and spatial distribution of central places and their hinterlands (after Christaller)

geographers began to develop similar kinds of theoretical arguments in relation to the role of large cities beyond the level of nation-states. Central place theories already provided an explanation for why the largest cities in nation-states were the only places that were the locations of some very specialist services and activities (such as national government and very specialist financial, legal or medical services), but as the world economy began to become globalized still further, it was clear that some cities fulfilled specialized functions at the international level.

In the 1970s and 1980s, this idea was developed into the theory of world cities, which suggests that cities such as London or New York were providing financial and other kinds of economic functions to the world economy. However, in the 1990s, as ongoing globalization became more apparent, a more developed version of this kind of theory was proposed in the form of a theory of global cities. Coined by Saskia Sassen (1991) in her book *The Global City*, the original argument was that leading cities in the global economy sat at the top of a global hierarchy of cities. Sassen identified

London, New York and Tokyo as the three globally most important cities, arguing that they held this position because they were the key locations where transnational firms had their head offices. In representing the command centres for an increasingly globalized world economy, these cities were also the only major locations for specialized business finance (i.e. investment banking and related activities) and other highly specialized producer services (corporate law or strategic management consultancy). As we have already discussed, the drivers for these functions to be concentrated in specific places have remained even in the era of global ICT, so these global cities represent a form of contemporary 'super-agglomeration' serving the global rather than the national level.

Urban geographers and others have developed these ideas extensively, arguing subsequently that these leading global cities are in fact at the head of an enormous network of cities around the planet that are interconnected as global city networks. In the second edition of her book, Sassen herself argues that global cities represent networked organizing hubs in the global economy of the 21st century. Geographical thinking has also been influenced by the arguments of the sociologist Manuel Castells, who suggests that the global city concept needs to be understood more like an (urban) process within globalization than simply a list of places (Castells 2009). The debate about global cities therefore centres increasingly on the issue of to what extent key functions in the global economy exist across this urban system in a networked form as opposed to being concentrated in a few specific 'global cities'. In this respect, urban geographers have evaluated both the relative importance of different cities with respect to their role in the global economy, and also mapped the international connections between them in order to assess how globalized they are (Taylor 2004). An important aspect of the global city thesis is that growing global interconnectedness between cities at different levels of importance means that their physical and social structures are changing due to global-scale influences. Cities such as London, New York or Frankfurt are thus argued to have increasing amounts in common with each other in terms of their labour markets and the factors that produce economic growth, more than they do with smaller cities within their respective nation-states (in the case of these example, cities such as Manchester, Chicago or Hamburg respectively).

Overall, geographers are interested in understanding how every city on Earth is increasingly bound into global-scale processes. The danger of course is that if all cities are now 'global cities', to what degree is the concept itself useful as an explanation of urban development?

THE POST-INDUSTRIAL CITY

One of the terms often used by urban geographers to describe cities since the later part of the 20th century is 'post-industrial'. In light of our discussion of the new economy and different Kondratieff waves of 'industrialization', this may seem a confusing description. Within some geographical and other social science work, however, it is used to refer to the nature of cities (largely in the global North) in the aftermath of the deindustrialization that began in the 1970s.

In that sense, cities are 'post-industrial' (literally 'after industry') insofar as they are no longer the locations of manufacturing or other traditional industries (for example, shipbuilding). What has now happened to many cities in the global North is in fact a diversification of the kinds of industry that are located there. Some still have manufacturing industries, but these represent a much smaller proportion of their GDPs and employ far fewer people. Many cities are increasingly dominated by service industries, and the wealthiest are often those that have a high proportion of these global-level business and financial services as well as the new creative industries discussed earlier. However, it is important to remember that human geographers use the term 'post-industrial city' rather loosely in relation to not just economic but social, political and cultural features of cities since the late 20th century. Post-industrial cities are those that have the characteristics of a wider post-industrial society. In that sense, the concept covers new kinds of urban forms, politics and culture. Overall, a more accurate term might be 'post-manufacturing cities' or 'new economy-based cities'.

URBAN REGENERATION

The concept of urban regeneration stems from the idea that it is possible to renew cities or areas of cities that have experienced deindustrialization and its associated problems – redundant and derelict buildings, population decline, environmental degradation, crime and decaying infrastructure. From a historical perspective, urban regeneration is very much an idea of the latter half of the 20th century. Its origins begin with the need to reconstruct many of the industrial cities of the global North after the destruction of the Second World War, but by the 1970s the question of urban renewal had become more pressing in light of deindustrialization. Early approaches to regeneration focused on the clearance of derelict buildings and the planning of new urban environments. However, regeneration projects from the 1950s and 1960s met with mixed success. For example, many cities in Europe constructed districts of high-rise residential tower blocks that reflected the modernist architectural ideas of those decades, only to find that the design of such buildings aggravated social problems such as crime. In recent decades, therefore, urban regeneration has been understood as a more wide-ranging process that needs to involve social and political dimensions. Urban geography today is very much concerned with the kind of broad definition of urban regeneration as 'a comprehensive and integrated vision that leads to the resolution of urban problems' and which 'seeks to bring about the lasting improvement in the economic, physical, social and environmental condition of an area' (Roberts and Sykes 1999).

One of the key issues, however, is the way in which urban regeneration is undertaken in practice, and how that is shaped by governments, private investment and local communities. The culture-led regeneration schemes that were discussed in Chapter 4 – waterfront developments in cities such as Baltimore, Liverpool or Bilbao – may all have been grounded in a similar set of strategies for making these cities more attractive as places to consume, but the manner in which regeneration was achieved varied considerably between them. Urban geographers are thus very interested in seeking to assess the degree to which urban regeneration projects in today's world are successful or not, and in providing a comparative assessment of what are the most effective ways of achieving urban renewal. An example would be work comparing the impact of hosting the Olympics on different cities, widely used as a tool to achieve urban regeneration (Rennie Short 2004). Consider how the London 2012 Olympics have an explicit goal of regenerating the poorer eastern part of the city, which has experienced decades of deindustrialization. In this respect, a key factor that urban geographers have increasingly focused on is seeking to understand how and why urban regeneration is successful is the nature of urban politics in different cities across the globe. It therefore makes sense to turn now to consider this issue more generally.

URBAN POLITICS

Urban geographers have made a significant contribution to wider debates within the social sciences concerned with the importance of urban politics. While much work within sociology, social history and political science has focused on the nature of political movements that have developed within cities, urban geography has been particularly concerned with the spatial form and consequences of urban political processes. Geographers see the process of urbanization as an inherently political process that plays out across urban spaces with uneven consequences. Central therefore to a geographical approach to urban politics is an emphasis on understanding how it shapes the development of cities, the global urban system and more generally affects the nature of wider global society. In the context of geographical theories of globalization, geographers are becoming increasingly interested in the relationships between the politics of cities across the planet (McCann and Ward 2011).

Several approaches to this within human geography are worth highlighting. First, a body of work within human geography since the 1960s has applied a Marxist approach to urban development, and argued that cities are important spaces where social processes such as class formation, identity formation and political conflict occur. Urban politics in this sense is important because it shapes the nature of wider society within nation-states and, in the 21st century, global society. Geographers in this respect have pointed to the way in which the spatial development of cities reflects class struggle in capitalist societies with, for example, theoretical attempts to understand why certain districts within cities develop as working or capitalist class areas.

Second, much work in urban geography has examined the nature of urban governance. In the last chapter we discussed how human geographers have conceptualized the way a global society of nation-states is governed, but a significant strand of work also looks at how cities are governed and what difference that makes to the way they develop in physical, economic and social ways. For example, Barnett and Lowe (2004) consider how cities are important spaces for the development of democracy and how urban politics shapes what kinds of ideas about citizenship exist in different places (see Chapter 6). A third, related, strand of geographical work is also interested in the key role that cities play in the geography of political parties and new social movements. There are many examples that could be offered here, but in terms of new political movements a good one is the new politics that have developed in cities across the globe around developing sustainability strategies in light of the problem of human-caused climate change. Urban sustainability taps into the wider global environmental movement but how specific strategies within different cities are developed is bound up with the particular politics of individual cities. A geographical approach to the question of how urban sustainability is achieved through political processes is crucial since it is impossible to understand all the factors involved without an understanding of the urban politics that forms across many scales – from the 'local' level of urban districts to that of 'global' environmental values and governance (Bulkeley 2005).

Fourth, a sizeable body of work in urban geography has been concerned with the significance of cities as places where informal politics and new forms of political resistance develop. Cities are the major places where political activity occurs. Think of the protests made by political groups in the symbolic spaces of capital cities, whether the US civil rights movement marching on the Mall in Washington DC in the 1960s, the anti-Vietnam War protests of the 1970s, or those against the Iraq War in 2003. Recent political revolutions in the Arab world in the last decade are equally good examples. Political resistance is intrinsically connected with urban places and human geographers are therefore interested in how cities are crucial places of political expression and transformation. This also extends to the politically contested nature of the urban built environment itself. Urban geographers are interested in how urban planning is a politicized process, and how the built spaces of cities become altered by the practices of the people who live in them. Good examples would be the way in which certain urban spaces become hijacked for different uses – pedestrian walkways and plazas being taken over by skateboarders or walls being taken over by graffiti artists.

SUMMARY

In this chapter we have considered:

- Economic geography's central concern with regions and regional economies, in particular how different industries have developed in certain regions within nation-states since the Industrial Revolution;
- The nature of agglomeration and clustering of firms and industries in today's global economy, and the arguments geographical thinking offers to explain this phenomenon;
- The different forms of knowledge and how knowledge is a key factor in understanding the nature of economic activity in the global economy;
- How innovation is one of the most important factors shaping economic success, and how industrial agglomeration aids innovation;
- Economic geographers' understanding of industrial development, and how different industry sectors such as manufacturing, services, creative industries and agriculture have changed in the last 50 years or so;
- How geographers have developed theories to understand the development and form of cities, urbanization as a process and urban politics including the idea of 'urban regeneration';
- The emergence of an increasingly globalized urban system, and the importance of global city networks in coordinating and controlling the complex 21st-century global economy.

FURTHER READING

Dicken, P. (2011) *Global Shift: Mapping the Changing Contours of the World Economy*. London: Sage.
This book, now in its 6th edition, remains the leading reference in economic geography for understanding industrial development, economic processes and the complex relationships between firms, industries and economies at all scales.

Mackinnon, D. and Cumbers, A. (2011) *An Introduction to Economic Geography* [2nd edition]. Harlow: Pearson Prentice Hall.

Look at Chapter 4 for a detailed account of how geographers have thought about industrial development, regions and clusters. Chapter 7 and 8 also provide good overviews of the debates surrounding the embeddedness of transnational firms and the service economy respectively.

Coe, N., Kelly, P. and Yeung, H. (2008) *Economic Geography: A Contemporary Introduction*. Oxford: Blackwell.

This book provides a good account of how economic geographers have engaged with debates about industrial development from a slightly different, more thematic perspective. Chapter 5 is particularly useful if you are interested in the debates regarding the relationship between technology and agglomeration.

Pacione, M. (2009) *Urban Geography: A Global Perspective*. London: Routledge.

Provides an excellent overview of how urban geography has developed its ideas in relation to cities and their place in today's world. As well as providing an historical and theoretical view, it is also strong on debates regarding the future of cities in relation to globalization and ongoing urbanization in the global South.

WEB RESOURCES

On global economic development, the Nobel prize-winning economist Paul Krugman has a good website and his ideas are influential in human geography: http://web.mit.edu/krugman/www/

Have a look at the website of the Globalization and World Cities group at Loughborough University: www.lboro.ac.uk/gawc/

6

PEOPLE, WORK, AND MOBILITY

This chapter examines how human geographers approach the study of population and changes to the composition and nature of societies including the idea of citizenship. It also addresses the movement of people in considering migration, along with new forms of mobility that have appeared in today's globalized world. The final part of the chapter then turns to think about work as it discusses the geographies of labour.

POPULATION AND DEMOGRAPHY

The study of population (known as demography) is another topic that has a long history in human geography. This is hardly a surprise, however, since the ways in which human populations on Earth change affects everything else: economy, environment, culture and politics. Human geographers have traditionally been most interested in three issues in relation to world population: the geographies of population growth, the factors driving this that have led to changes in the birth rate and death rate, and the consequence of population change for nation-states, regions and the global environment.

POPULATION GROWTH

It is impossible to measure total global population accurately. Most countries in the world today conduct a census of their populations periodically (normally every ten years or so), and other sources of data provide the basis for developing reasonable estimates of population. The World Bank estimates that in 2009, the global population had reached around 6.8 billion planet-wide, and based on a current average growth rate of 1.2 per cent per annum, it will double in 58 more years or so (i.e. by the year 2067). Predicting future rates of population growth is, however, difficult. In fact, it is widely expected that the rate of growth will decline worldwide and that therefore the total global population will not be as high as this figure. Nevertheless, if you consider that in 1960, the estimated total global population was only 3 billion, you can appreciate the significance of population growth as an issue facing the planet and everyone on it. Many would argue that it is one of the biggest challenges facing human society.

Historically, before the 18th century, total global population had been growing on average only very slowly, and taking a long view of history back to Roman times, populations in certain regions of the world had also decreased rather than increased. However, with the Industrial Revolution and the emergence of modern capitalist societies, the rate of population growth increased dramatically. Population growth during the 20th century became a more or less global-scale norm, although certain places and regions did suffer decline at times. It is important to realize, therefore, that much of the growth in total population in human history is something that happened during the 20th century. While it is true that there was significant population growth in some of the industrial countries of the global North in the 19th century, in the first couple of decades of the 20th century the overall global average rate of population growth was only around 0.5 per cent. During the 1920s, however, this rate more than doubled at the global level to around 1 per cent until the late 1940s. By the 1950s, there was a further increase and the global rate of population increase reached 2 per cent per annum by the mid-1960s. At this rate of change, the actual total global population doubles every 35 years. Since the 1960s, however, the growth rate has fallen back to today's 1.2 per cent per annum, but given that global population has already increased enormously in absolute terms over the last century, that still produces an annual growth of an extra 35 million people per annum. Even the most optimistic projections for further decline in the rate of growth suggest that planet Earth will have a population of 9 billion by around 2050.

In simple terms, growth in human population reflects a higher rate of births than deaths. If the number of children born each year in a given country or region exceeds the number of people that die, then population increases. Of course, what is behind these changes in the rate of births and deaths is the important thing to understand. There are many factors, related to the changing nature of human societies, economies, technology and medicine in different parts of the world. Human geography is thus particularly concerned with understanding the unevenness of a growing global population, what factors are contributing in different regions and what implications these have for societies and economies. Looking at headline figures for global population growth reveals little about this. We therefore need to describe some generalized features of the geography of growth over the last two centuries.

First, in broad terms during the 19th century, population growth was concentrated in Western Europe and a few other areas where industrialization was taking place. It was only in the 20th century that the populations of less developed regions grew rapidly, and most significantly since the Second World War. A second issue is that, since the 1950s, the rate of population growth in countries of the global South (including the faster-developing economies of Asia) has increased. Populations in these regions of the world have increased from around 1 billion in 1950 to more than 5 billion now. Conversely, the populations of the wealthier advanced industrial economies have remained more or less stable, with little growth. Roughly 1 billion of the Earth's total population live in these economies of the global North. The overall point is that population growth over the last half century or more has mainly occurred in the global South. Finally, it is worth noting that the higher proportion of the growth to date has been in Asia but that over the next 40 years it is populations in Africa rather than Asia that are expected to increase most rapidly.

There are clearly huge implications for a continual rise in the number of human beings on Earth. Geographers have in the past been concerned with the obvious question of how many people the planet can support in terms of straightforward issues of resources (known sometimes through an ecological population term, 'carrying capacity'). At a basic level, people need food, water and shelter. Even as recently as the mid-1960s famine in China killed more than 20 million people and later millions more died in famines in Africa in the 1980s and 1990s. However, modern famine is more about the availability or production of food in a certain region than an absolute lack of it on Earth. Most recent analyses suggest that, in today's era of modern industrial agriculture, the planet can produce enough food to feed its current population and can continue to produce more food as the population rises. However, what impact this will have on the Earth's environment, including its climate and biodiversity, is worrying. Population growth, along with the geography of where it happens, thus remains a major issue at the centre of human geographical debates about sustainable development

We also need to think about how human geographers have made use of various theories to understand population growth and its relationship to wider changes in global society. In this respect, geographers have made much use of a well-known model to understand the way in which population change has changed within countries as they develop: the Demographic Transition Model (see box and Figure 6.1). It was developed in the 1920s and is based on the historical experience of the advanced industrial countries in the global North. The model provides a means of understanding how the timing of changes to the birth and death rates is related to the rate of population growth. In essence it suggests that as countries industrialize and become modernized they at first experience a period of rapid population growth before this slows down and the level of population becomes stable (as it is in many advanced industrial countries today). However, human geographers have reassessed the validity of this model as a way of understanding how less developed countries in the world today experience population change. Geographical work has put much effort into understanding how demographic transition has varied and continues to vary between different countries in the global South, and looks for a more sophisticated understanding of the many factors that affect population change than are captured by this model.

THE DEMOGRAPHIC TRANSITION MODEL

One of the most influential ideas about the relationship between development and population in human geography is known as the Demographic Transition Model. Developed in 1929 by an American demographer, Warren Thompson, it describes five stages of a relationship between the total population and its level of economic (and social) development based on changes in the **birth rate** and **death rate**. The model is based on observations of changes to these rates over a 200-year period in Western societies as they experienced industrialization. It describes three main stages that countries have passed through in their demographic histories (see Figure 6.1).

In stage one, a country has both a high birth rate and a high death rate. Health care is basic and people die of diseases that could be easily cured by today's medical standards. People have a lot of children and large families are the norm to ensure enough children survive long enough to look after their parents in old age. With both the birth and death rate relatively high, population does not increase much during this stage. For Western countries, this stage corresponds to the period prior to the development of industrial society in the 19th century.

Stage two of the model is characterized by a declining death rate as sanitation and medicine improve as a country develops. Fewer people die from easily curable diseases. However, in this stage, the birth rate remains high, perhaps for cultural reasons or because economic growth means there is demand for more labour. The important point is that in this stage rapid population growth occurs, caused by the gap between the birth and death rates. In the real world, many developing countries experienced this kind of situation at some point during the 20th century.

The third stage of the model sees birth rates begin to fall towards the level of the lower death rate. In this stage, population is still rising but at a slower rate than in stage two. The reasons for this are multiple. To begin with, increased economic prosperity among the population in a country along with the development of modern pension systems means people have to rely less on their children to support them in old age. Also important is increased urbanization, since people in cities have less space in which to bring up large families. In addition, however, the declining birth rate is attributed to a range of broader societal reasons: women have greater educational and employment opportunities, the increased availability of contraception, the fact that many women have their first child later in life and changing cultural attitudes that mean having a large number of children is no longer seen as the norm.

These latter factors produce the later fourth stage of the model, which is argued to be that typical of most advanced industrial countries since at least the mid-20th century: both a low birth rate and a low death rate, which results in a stable level of population within a country. In some developed countries, however, birth rates have fallen below death rates, which leads to a situation of population decline over time (see the section on geographical variations and population crises below).

GEOGRAPHICAL VARIATIONS AND POPULATION CRISES

Aside from the question of growth, work within population geography has also been concerned with a range of geographical variations in population between different states and regions. Much of this is concerned with the factors that shape existing differences in the structure and composition of populations in different countries, and the impact of these population structures on future social and economic development. There is only space here to discuss a few examples from a large body of work within human geography, but at least two are worth highlighting.

Regarding variations in population structure first, different countries around the world have different demographic structures in terms of, for example, the proportion of their populations in different age groupings or the proportion of men to women in different age groupings. These can be seen in the population 'pyramid' graphs shown in Figure 6.2, which reveal a country's demographic structure. While providing some useful generalizations, the apparent 'stage' of a country's population in the demographic transition model provides only a limited understanding of actual demographic structure. For example, some countries or regions have a high proportion of young people, or an underrepresentation of men of working age. In particular, within many countries these factors vary

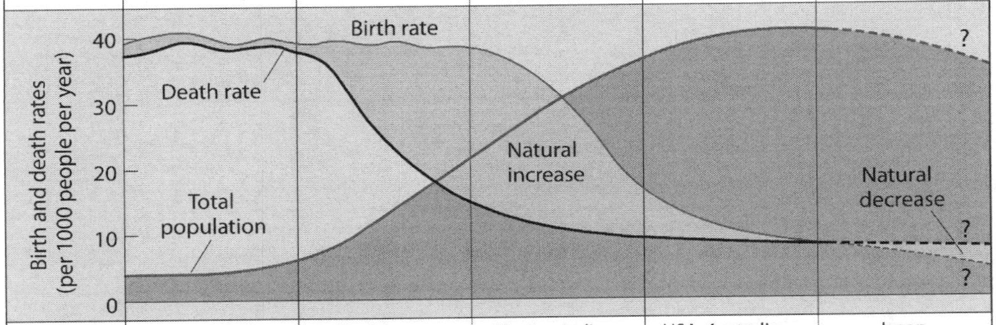

Stage	1. High stationary	2. Early expanding	3. Late expanding	4. Low stationary	5 ? Declining?
Examples	A few remote traditional communities	Chad, Laos, Afghanistan	Mexico, India, South Africa	USA, Australia, most of Europe	Japan, Netherlands
Birth rate	High	High	Falling	Low	Very low
Death rate	High	Falls rapidly	Falls more slowly	Low	Low
Natural increase	Stable or slow increase	Very rapid increase	Increase slows down	Stable or slow increase	Slow decrease

Figure 6.1 The demographic transition model

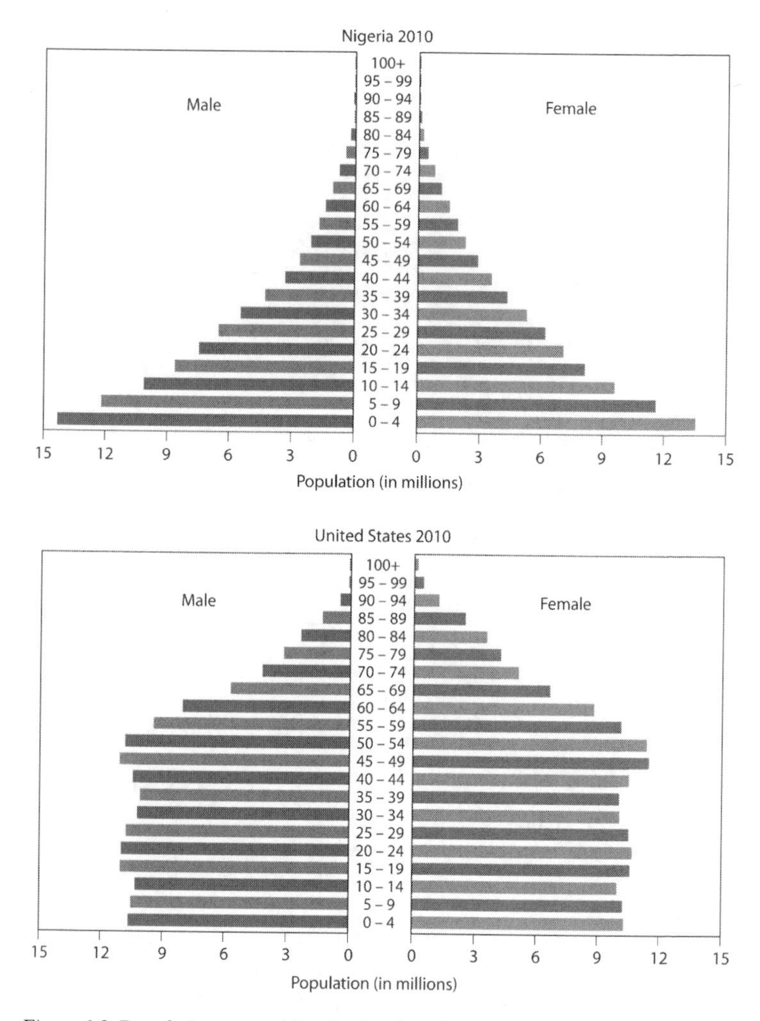

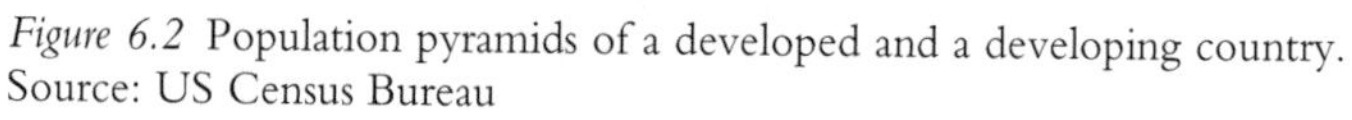

Figure 6.2 Population pyramids of a developed and a developing country.
Source: US Census Bureau

substantially between different ethnic groups within the total population. In that sense, there are usually country- and region-specific factors behind differences in population structure that need examining in more depth. Religious or cultural contexts, for example, can be a factor, or the inward or outward migration of certain groups within the population. Such differences have significant impacts on real economies and on the nature of societies and place.

Human geographers are thus often concerned with digging into the complexity of population structures within a country that provides some indication of likely future changes and issues in relation to population change. Many countries in the global North, for example, have a growing proportion of young people in certain ethnic groups. In Western European countries such as the UK and France, a higher proportion of young people are from ethnic communities as a result of colonization in Asia and the Middle East. (We will discuss the issue of migration in more detail shortly.) The growing representation of young people in these ethnic groups within these populations reflects several decades of a higher birth rate among these groups within the country's population. Similarly, the US has a growing proportion of younger people of Hispanic ethnicity reflecting decades of in-migration from Mexico and other Latin American countries. Over time, this is likely to have several effects in terms of increasing cultural diversity within a country or affecting the nature of domestic politics. In that sense, a geographical understanding of the diversity of population structures and of future changes to those structures provides an important basis for an understanding of the major forces shaping different societies around the world.

Also, population geographers have been very much concerned with population crises facing certain countries in the world. One important example that has received a lot of geographical attention is the impact of HIV/AIDS on many countries within Africa. The human immunodeficiency virus (HIV), which can lead to acquired immunodeficiency syndrome (AIDS), was first identified in the mid-1980s. The disease is still without a complete cure, although a range of drugs have been developed which inhibit its development. Importantly, however, the impact of HIV/AIDS on the global population has been highly uneven within some regions over the last few decades. In sub-Saharan African countries in particular, the effect of HIV/AIDS on populations has been dramatic. Consider that, for many countries in Western Europe, rates of HIV infection have never exceeded 1 per cent of the population and contrast that with some countries in Africa that have seen infection rates of over

20 per cent and as high as 50 per cent in the last two decades – countries such as Botswana and Zimbabwe. There are multiple reasons for this but some important factors are: the high cost of the drugs used to treat HIV/AIDS, which were unaffordable in the low-income countries of sub-Saharan Africa; the lack of resources to educate people on the danger; as well as societal factors such as the position of women in traditional African societies.

THE AIDS CRISIS IN BOTSWANA

Until the mid-1980s, the southern African country of Botswana was one of the more successful developing countries in terms of increasing life expectancy and the health of its population. Botswana had seen its life expectancy rise during the 20th century to around 65 years by the mid-1980s. However, the impact of HIV/AIDS on the population in the last 30 years or so has been traumatic. HIV/AIDS had reduced life expectancy in Botswana to 34 years by 2006, with the death rate exceeding the birth rate. It is estimated that nearly 100,000 children have lost one parent to the disease in the country at present (UN AIDS 2010). Since 2005, the government, with other aid agencies, has run an increasingly successful prevention and treatment programme but the situation still represents a population crisis. With the death rate still exceeding the birth rate, Botswana has a shrinking total population. Socially and economically the impact of HIV/AIDS has also been devastating. There is a shrinking workforce to support economic activity, and generations of young people grow up without both parents.

While in recent years some sub-Saharan African countries have been successful in reducing rates of infection, in 2010 it was estimated that there are still around 23 million people are living in the region with HIV/AIDS, with 1.8 million becoming infected and 1.3 million people dying of the disease in 2009. In the last two decades, it is no exaggeration to state that this has produced population crises in some countries (see box on AIDS in Botswana).

However, the nature of population crises is not restricted to the more dramatic effects of diseases such as HIV/AIDS. Some population crises have less obvious causes than disease and are bound up with multiple factors that are having – and are likely to continue to have – a significant impact on countries and regions around the world. Examples that concern human geographers are the impact of ageing populations on the future economic and social viability of different countries and regions (see box on ageing in the global North) or the impact of significant inward or outward migration of certain groups within populations (such as people of working age) from or to different countries. Any demographic transformation that leads to an imbalanced population structure can have significant negative consequences for a society, whether in terms of prosperity, the continued reproduction of the population or equally in cultural and political terms.

A CRISIS OF AGEING IN THE GLOBAL NORTH?

Human geographers have been particularly interested in one of the big population challenges facing the wealthier countries in the global North – populations that are getting older. An ageing population is a characteristic of countries in the late stage of the demographic transition model where birth and death rates have converged, with in some cases the birth rate dropping below the death rate. Over the last couple of decades, this trend is evident in the UK, France, the Netherlands, Scandinavia and also in Japan. The consequence in this situation is that the average age of the population increases, with fewer and fewer children being born.

Consider for example the situation in Japan, which is currently the fastest-ageing country on Earth. It is estimated that by 2020 there will be three people of pensionable age in the population for every child, and even before then, one in six people will be over the age of 80. Furthermore, within a few years if the current trend continues, Japan's population will be shrinking by around 1 million people per year. Such a situation presents a range of very serious problems for Japanese society and the economy, as it does in other parts of the global North: too few people of working age to keep the economy productive as well as the growing burden on healthcare and pension systems. The problems of an ageing population can of course be addressed through the in-migration of younger people into a country or region, but given the politicized nature of immigration this is not always an easy or unproblematic solution.

CITIZENSHIP

Almost everyone on the planet in the world of the 21st century is (notionally at least) a citizen of one nation-state or another. Defining citizenship is straightforward from the perspective of political science: 'The terms of membership of a political unit [usually a nation-state] which secures certain rights and privileges to those who fulfil certain obligations' (Smith 2009). Understandings of citizenship often identify in particular three forms of rights that citizens have: *civil* or *legal* (for example, freedom of speech, assembly, movement), *political* (voting, holding political office) and *social* (entitlement to social security and state benefits) (Muir 1997). Importantly, it is essentially a conception of the conditions for membership, participation and rights in a community that is associated *with a given territory* (Smith 2009). This last point is the crucial reason why human geographers have been so interested in the concept, since an individual has to be a citizen of a given place. Furthermore, that community to which they belong or participate in is situated in a certain place. It is an inherently geographical idea.

It is very likely therefore that as you read this you are a citizen of a country. You will either have a passport issued by that country or the right to receive one, and you are governed by that country's laws and requirements. As a citizen, you are also probably a political participant in your nation-state's existence. In much of the world, that means voting in elections, but even in non-democratic countries where not everyone (or anyone) votes, citizens may have to be part of political institutions or governing organizations. Many countries require citizens to provide certain services to the nation-state – some demand that young people undertake a period of military or civic service and many require their citizens to volunteer their time to serve as jurors in the judicial system. Conversely, nation-states have a responsibility to look after and attend to the needs of their citizens – most notably to protect them. Citizenship is therefore a conceptual relationship of membership between individual people and a nation-state.

Interestingly, the universality (if not consistency) of the concept that nation-states are composed (in human terms) of citizens means that in the 21st-century global political system we live in, ideas of citizenship are reinforced at the supranational scale. International law and institutions such as the United Nations ensure the concept of citizenship exists above the scale of nation-states. This is in the background of international politics for much of the time but becomes clearly visible around the global politics of war and conflicts in today's world. When nation-states enter into civil war or act in a totalitarian manner, international actions by the UN or military organizations such as NATO are often justified by and grounded in arguments concerning the abuse by a state of its citizens. Examples would include military interventions in the Yugoslav war of the 1990s, the 2003 war in Iraq and the NATO intervention in the civil war in Libya in 2011. In this way, there is a developing concept of global citizenship that sees individuals with rights and responsibilities as global citizens beyond their relationship with any given nation-state.

Human geographers have become increasing interested in the last issue, but more widely there is a lot of geographical work concerned with differences between concepts and practices of citizenship in different parts of the world. Much work has also in particular focused on what has been termed the 'spaces of citizenship', with geographers pointing out that the spaces of citizenship within nation-states are not 'straightforwardly inclusionary' (Painter and Philo 1995). In fact, whilst the language of citizenship is that of inclusion, it is an exclusionary practice. Social geographers point out that historically only select groups have been entitled to citizenship (Valentine 2001). That is to say, despite the ideal of citizens of any given state being equal members and participants in that community, in reality in many countries populations are divided along longstanding gender, sexuality, religious, cultural, racial and ethnic lines. Many groups have only seen a slow extension of the rights of citizenship over time and through political struggles. Consider, for example, the experience of the women's liberation movement in the global North at the start of the 20th century or the American civil rights movement after the Second World War. Geographical thinking has thus argued that the ideal of citizenship does not exist in reality.

Citizenship therefore exists as an uneven physical, socio-cultural and political space accessible by different people to varying degrees. In some cases, there are formalized divisions in concepts of citizenship with certain communities living within nation-states denied the rights of full citizenship. An often cited example is the case of guest workers that came to work in West Germany (before German unification in 1990) after the Second World War (one of the largest groups being Turkish). These workers were intended to return to their 'home' countries but most never did. However, for a long time they existed as second-class 'non-citizens' in Germany, denied social rights and other benefits. They were also historically restricted in the extent to which they could move around Germany, as well as being confined to specific places and types of accommodation. In this sense, these workers were caught up in all kinds of physical and social exclusionary spaces in relation to citizenship. Another example is the limited rights offered to lesbians and gay men, with social geographers pointing to their exclusion from full citizenship (Browne et al. 2009). There are therefore many dimensions to exclusionary spaces of citizenship. One widely examined aspect is the way heterosexual men and women can acquire citizenship though marriage or legal partnership in most countries of the world whereas very few countries offer the same opportunity to lesbians or gay men (Valentine 1995; 2001).

However, geographers are also interested in less formalized kinds of exclusion from citizenship. Social geographers' work on sexuality has demonstrated in this respect how the 'spaces of sexual citizenship' are exclusionary in multiple ways that involve both physical and material spaces (buildings, cities, places) and social spaces (social groups, communities and networks). In many liberal countries of the global North – in Western Europe and North America – the argument here is that it is possible to be gay only in certain specific places and spaces. Being visibly gay in northern Europe may be easy, acceptable and comfortable in a gay bar or club but the same is not true for a gay man at a football match or in certain urban districts at certain times of day. Geographers have argued that this reveals the uneven spaces of what might be called 'everyday citizenship'.

Other work been concerned with understanding how different groups reshape ideas of their own citizenship around alternative spaces (in physical, social and virtual ways). For example, geographers have looked at the importance of gay neighbourhoods in cities such as London, Amsterdam or San Francisco in developing senses of citizenship (Binnie and Valentine 1999; Browne 2006).

This has shifted geographical understandings of citizenship beyond the narrow idea of rights and responsibilities associated with national community membership. Rather, geographical thinking sees citizenship increasingly broadly as a set of practices that occur in other spaces – the home, the neighbourhood or rural areas (Parker 2002) – and which are related to multiple senses of belonging by people to different communities. With the ongoing development of virtual communities in cyberspace, such concerns are already shifting further towards consideration of virtual networks of citizenship through social media such as Facebook, MySpace and other online spaces.

DIASPORA

The concept of the diaspora is used increasingly by human geographers to describe the globally scattered populations of members of certain ethnic groups. The word 'diaspora' itself simply refers to that kind of spatial distribution: it is the kind of scattered spatial pattern you would get if you dipped a paintbrush into a can of paint and flicked it against a wall. Diasporic populations or groups of people that are generally referred to in this way are people of Jewish ethnic backgrounds, people of black African descent and people of ethnic Chinese origin living around the world. Diasporic communities are therefore the consequence of past patterns of migration going back many centuries, and in particular where people have moved to many different places and regions around the planet. The idea of diasporic communities takes the concept a stage further, however, from this and is – to a degree – more debatable. Some commentators, thinkers and politicians have argued that different diasporic groups correspond to what is effectively a scattered nation with a common identity, usually based around an ethnic origin and culture. Such an idea is controversial, however, since many diasporic communities are made up of different groups that have also developed significant cultural distinctiveness over time.

Two examples are particularly useful and have widely concerned human geographers. The first is the global Jewish diaspora. Focusing on religion and cultural commonality, it is quite easy to argue that a coherent Jewish diaspora does exist, with various people of Jewish descent around the globe identifying themselves strongly with that sense of scattered community. The history of the development of this diaspora goes back many centuries, and corresponds to complex patterns of migration that occurred for a variety of reasons, including the historical persecution of Jews in Europe. While there are different strands to the Jewish religion, people in this diaspora continue to share much commonality in ethnic, religious, cultural and linguistic terms. The nation-state of Israel was created in 1947 in the Middle East partly on the basis that the Jewish diasporic community had no national homeland and needed one in the modern world.

However, other concepts or imaginaries of diasporic communities are less well defined. This is well illustrated by an example in the form of the black African diaspora in the countries around the Atlantic Ocean. Termed the 'Black Atlantic', this diasporic community consists of people of black African descent now living in Europe, the Americas and Caribbean. The ancestry of these people is largely connected to the European slave trade between the 16th and early 19th centuries that forcibly relocated black Africans from the African continent and scattered them on both sides of the Atlantic Ocean (see box). Such a diasporic community undoubtedly represents a looser sense of the concept since many of the people involved, while of black African ancestry, may have much less in common in linguistic, cultural or religious terms than people considered part of the Jewish diaspora. The idea of a diasporic community is thus a variable one, and geographical thinking has been very much concerned to consider to what extent a given supposed community actually represents a coherent group of people with a shared identity, culture and understanding.

THE 'BLACK ATLANTIC'

In his 1992 book *The Black Atlantic*, the cultural theorist Paul Gilroy argues that people in many countries of Europe and North America are united in a diasporic community based on their common ethnic ancestry and, equally importantly, on the historical experience of slavery and racism. The identity of the black diaspora that is scattered across Europe and the New World has emerged, he suggests, as the result of an ongoing process of travel and exchange across the Atlantic. The senses of identity that have resulted are a consequence of an ongoing process of defining the African black diaspora in relation to European modernity. The book itself is therefore an argument for the active generation of a certain kind of diasporic transnational identity that recognizes how the history of slavery is bound up with the dominance of Western culture over black African culture.

Such an argument is similar and related to the work of Edward Said insofar as it engages with historical questions of representation and imagined identities (see Chapter 4). Gilroy contends that a shared experience of 'terror' lies at the heart of black diasporic communities all across the Atlantic. He identifies this as the 'root cause' of transnational black identity and suggests that for 150 years black intellectuals have travelled and worked in a transnational frame that has more or less excluded anything but a weak or very limited association with their country of origin. The point is that the senses of identity shared by the black diaspora are inseparable from a complex history of African diasporic culture, which is itself transnational in nature and has been for a long time. Such a set of arguments has been important for human geographers in attempts to better understand diasporic identities and the way they have developed in time and space. However, they are not unproblematic. Critics of the 'Black Atlantic' concept argue that this community is in fact fractured and incoherent and that there is little genuine shared sense of identity or scope for developing the imaginary of a black diasporic nation.

MOBILITY

For a long time, questions of the movement of people in human geography focused purely on migration. We will consider this shortly, but first it is important to emphasize that in the last couple of decades the subject has become increasingly concerned with differentiating migration from other forms of mobility that have become important in the last century or so. People are moving with increasing frequency and in all kinds of ways and, over the last 50 years or so, this has had enormous implications for economies, societies and people's daily lives.

Geographers have become caught up in a wider debate in the social sciences about the crucial importance of mobility in today's world. In broad terms, this is very easy to establish factually. There are more than 200 million migrants living in the world today, with the global tourist industry being worth nearly $1 billion and, for example, more than 675 million people taking a flight in the United States during 2008 alone (Adey 2009). The scale of people movement on Earth today is staggering. Yet such figures also perhaps conceal the less dramatic or noticeable mobility that is present in everyday life. Think of all the people commuting daily to work, children travelling to school or people going to buy food in shops. However, in theoretical terms human geographers have argued that mobility is even more ubiquitous. Not only people move but also objects, images, ideas and knowledge, which we experience constantly. Geographers have thus also become interested not just in how people move, but in how the mobility of other things affects people all the time. What happens in a given fixed place such as a city is often entirely dependent on other things being in motion. An airport exists as a meaningful place because of the constant movement of people, planes and numerous other objects in and out of it. The same applies to all places, whether offices, factories, homes or shopping malls. The argument therefore is that mobility is everywhere insofar as it is something we do and experience all the time. However, there are many different kinds of mobility, with some forms being more or less extensive or at differing speeds.

MIGRATION AND IMMIGRATION

Migration is a topic rarely out of the headlines in the global media. If you live in a country in the global North – whether that is North America, Europe or Australasia – you are likely to have encountered vigorous political debates in your country about whether migration is desirable. Usually this is about whether or not people from other parts of the world should be welcomed or prevented from coming to live in your country, and most debate is concerned with whether migration creates a benefit or a burden on the host country.

However, the concept of migration has a much broader usage in human geography. In simple terms, migration refers to the movement of people, and there are many different kinds of reasons that lead people to move from one part of the world to another.

Geographers and other social scientists often talk about 'push' factors that cause people to move, such as political instability, war or enforced migration. Equally, there are argued to be 'pull' factors as well, which include employment opportunities or the attractiveness of the environment in a given area. However, migration is another of those concepts that seems relatively straightforward but in fact proves more difficult to define upon closer consideration. For a start, it is often used interchangeably with the term 'immigration', although they are not strictly speaking identical – immigration refers more specifically to people moving across national borders in the global system of nation-states (see the discussion of the state in Chapter 4). Moreover, even in simple terms, not all forms of human movement equate to either term. Tourism is an obvious example we will consider later in this chapter, but another mentioned already would be the daily commuting by workers in large global cities such as London, New York or Tokyo, which does not fulfil general understandings of 'migration'. Yet if those workers spend part of their time living in those cities during the week before returning to another home at the weekend, then we might regard that as a form of temporary migration. The key issue is that migration is actually a specific form of mobility that needs to be distinguished from others, and the distinction between what represents 'migration' as opposed to other forms of mobility is blurred. How far a person moves and how long they stay in a place tends to define whether they are regarded as a migrant as opposed to a commuter, traveller or tourist.

Human geographers therefore tend to categorize movement that does correspond to migration by using criteria centred on time and space. Regarding the temporality of migration – that is, how patterns of migration vary over time – much work has sought to analyse what might be understood to be 'permanent migration'. For example, in the modern period of recent centuries, many hundreds of thousands of Europeans moved to a new part of the world to live and largely remained there. Consider the many people from the British Isles who moved permanently to North America, Australia and parts of Africa during the period of the British Empire. Many of the people who perished when the ocean liner *Titanic* sank in 1912 were similar migrants on their way to New York. Given the transport technologies and costs of long-distance travel until the mid-20th century, people mainly moved to the 'new world' to settle and live the rest of their lives there. Few returned to their home countries. Geographers have historically done much analysis of past permanent migration patterns going back centuries. However, in today's world, what corresponds to a permanent form of migration is much less clear. People are much more able to return to their country of birth with modern forms of international travel, and it has become rare for individuals to migrate from their home country to another and then stay there for the remainder of their lives. Unlike a mid-19th century emigrant from Europe, you can return to your home country on a plane quite easily and comparatively cheaply from anywhere on the planet within a period of a day or so. That does not mean that permanent migration no longer occurs, far from it, but it is much harder to define, map and measure.

A second way of categorizing migration concerns the reason or motivation behind it. Why do people migrate? Human geographers distinguish between voluntary and forced migration, and economic as opposed to political reasons for moving. People who leave their home country in order to find a job or seek a better quality of life elsewhere do so voluntarily as economic migrants, whereas those who flee a civil war or a famine move because they are being forced to and are refugees.

Finally, a third important distinction is between legal and illegal migration. In today's global system of nation-states, whether an individual is permitted to move to another part of the world and stay there is bound up with a range of laws and policies that vary between different nation-states and regional super-states such as the European Union. For political migrants, international law gives people the right to seek asylum (see below). However, the legality of an economic migrant depends largely on the policies and laws within certain states. Some countries actively seek and welcome inward migration to fill gaps in their labour markets, while others seek to prevent it. If you as a European go to work in the United States, for example, you will need to get the required permit that entitles you to work there legally. However, many economic migrants in today's world move without the legal sanction of the state they move to. Current examples include the continued flow of north Africans sailing small boats across the Mediterranean to enter the European Union, or Mexicans crossing the southern border of the United States. Migration in today's world is thus often a highly politicized issue, frequently making news headlines.

Beyond examining these kinds of categorizations, human geographers are keen to understand how the nature of migration itself in our 21st-century world is changing. At least four important trends have been identified which are the focus of much current work on the topic (Castles and Miller 2003). The first is that, not surprisingly, migration as a phenomenon is becoming more globalized. More countries and regions of the world are affected by it, and there is growing diversity in terms of the places where people migrate from and to. Second, the process of migration is accelerating with the volume of movements increasing worldwide at the moment. A third trend is that migration is becoming differentiated insofar as in the world today there are many more types of movement between countries. That is, if arguably permanent migration was the dominant form at some points in the past, it is now combined with the economic migration of skilled labour, refugees, students, retirees and others. Finally, and perhaps less obviously, migration is becoming feminized. In short, more and more of the people moving are women. Some of these are women joining male partners who have migrated, but there is a growing number of women who are migrants in their own right. An example of this would be the thousands of Indonesian women who have left to work as economic migrants such as care workers, cleaners or household maids in wealthier places such as Singapore or the United States.

REFUGEES AND ASYLUM

The concept of a refugee in the contemporary world is defined in international law in the Geneva Convention of 1951. The convention defines a refugee as an individual who has migrated because of a 'well-founded fear of being persecuted for reasons of race, religion, nationality, membership of a particular social group or political opinion', and is now outside his or her country of nationality. A further element emphasizes that a refugee is afraid to return to their country of origin because of that fear. The concept has thus changed over time, and really has only had its current meaning in international law since the late 20th century. The

Office of the United Nations High Commission for Refugees (UNHCR) estimated that there were at least 15.2 million refugees worldwide in 2009 (UNHCR 2010). The most common reason is war and conflict, but other reasons include famine and persecution (see box below).

Closely linked to this idea is that of 'asylum'. Although it has its origins in the idea of religious asylum in medieval Europe, in the modern world someone who 'seeks asylum' does so to seek protection from a country other than the one of their nationality. This concept of asylum exists as a right under the 1951 UN convention relating to the status of refugees. It was further developed in a 1967 UN Protocol. It is important to realize, therefore, that prior to this point there was no such legal status internationally for asylum-seekers and few migrants claimed such a status. Seeking asylum is thus an aspect of migration that has only become possible recently as a global-scale system of nation-states has developed that is governed by a range of supranational organizations like the UN (see the discussion of governance in Chapter 4).

In recent decades, the issue of asylum has become an increasingly politicized one because of the particular international geography of asylum applications. In 2008, there were around 350,000 asylum requests in the 34 wealthier countries that form the **Organisation for Economic Co-operation and Development** (OECD) (that is, loosely, in the global North), and this represents an ongoing trend of increase. However, of these just five countries received nearly half of all applications: the UK, the US, France, Canada and Italy each having more than 30,000 applications. In 2009, the UN estimated that the global total for asylum applications was just short of 1 million people. Thus, globally there has been a dramatic increase in the number of international asylum seekers since the late 1980s but there is a very distinct geography to where applications are being made. Reflecting this, over the last couple of decades, the European Union has sought to develop a common policy approach to deal with this significant increase in asylum applications.

While some politicians in high-application-level countries such as the UK and France have argued that many of these asylum applications are unjustified, in reality the drivers behind the increase in asylum applications are more complex. At least three factors are important. First, increased asylum applications worldwide are undoubtedly also related to wider processes of globalization that have led to more and more people across the globe being aware of the possibility of claiming asylum, the international laws and institutions that support it and the scope for increased global mobility (the growing capacity and falling cost of air travel) that makes it easier for them to move (see the section on mobility). A second factor is the continual occurrence of wars and conflicts around the globe, as well as oppressive nation-state governments causing people to flee. In the last few decades, there has been no shortage of conflict producing significant exoduses of people who have sought refuge from persecution. Examples include the various wars in the former state of Yugoslavia in the 1990s and more recently the repressive regimes in the Middle East and Asia (to a lesser extent). Relatively large numbers in Europe and North America have come from Iran, Iraq and Afghanistan, prior to the 2002 war.

Third, a wider knowledge of the international legal status of asylum led some people who wished to be economic migrants to claim such status. In the UK, France and Germany such migrants were dubbed 'bogus' asylum seekers from certain political perspectives, but in reality, determining whether an individual is a true asylum seeker is a complex and often subjective process. It is inevitably a matter of judgement and speculation for any country to assess the likelihood that people claiming to flee political persecution from oppressive states will actually be persecuted if they are returned to their country of origin.

THE REFUGEE CRISIS IN DARFUR, SOUTHERN SUDAN

Since 2003, a civil war in the southern region of Sudan in east Africa has forced hundreds of thousands of people to flee as refugees; the UN estimates that more than 300,000 people have been killed. The war started when a rebel army, the Sudan Liberation Army (SLA) and another group, the Justice and Equality Movement, started attacking government targets and accusing the Sudanese government of oppressing black Africans and favouring people of Arab descent. The conflict focuses on the region of Darfur – an area that has had a long history of political tension around land and grazing rights.

Not surprisingly many people have sought to flee the conflict. This has happened at several scales. By the far the greatest movement of people was the exodus from the towns in the region. By early 2010, it was estimated that around 2.7 million people had fled urban areas and were living in temporary camps nearby. These people are not conventional international refugees, not having moved across a border, but their situation is comparable. However, at the international scale, it also estimated that at least 200,000 conventional refugees had crossed the border into the (poor) neighbouring African country of Chad. These people moved to refugee camps along a roughly 500km stretch of the border and suffered attacks from rebel forces from the Sudanese side of the border.

Darfur's refugee crisis illustrates the complex nature of the nature of what is meant by a 'refugee' in today's world as well as the difficulties in tackling the humanitarian crises that refugee movements of this scale often produce (especially in the global South). With an ongoing civil war, providing these people with their basic needs of foods, water, shelter and safety is challenging – let alone any longer-term solution to re-establishing them in a permanent home.

TRANSNATIONALISM

Another concept that links debates in the last section around mobility, citizenship and diasporas is the concept of transnationalism. Literally, the word 'transnational' means across nations and the idea of the transnational firm should be familiar from Chapter 2. Human geographers have, however, used the word in several ways to try to capture the complex connections in social life that exist in an increasingly globalized world. What is important about the idea of transnationalism is that it bridges static ideas of migration (people moving to another place) with the dynamic maintenance of connections between places at the global scale. These connections can relate to all aspects of life – social, cultural, political and economic – and represent a new phenomenon in a globalized world because previously these aspects of life largely existed below the national scale (for example, senses of identity). Transnationalism thus has many potential applications from a geographical perspective in theorizing global-scale interconnections. However, at least three areas of work in particular have made use of the concept in recent years.

The first relates to our earlier discussion of diasporas. Political geographers have used the concept of transnationalism to understand political movements, identities and organizations that exist beyond the scale of the nation-state. Some of the diasporas discussed earlier are examples in this respect. The development and maintenance of a common sense of identity in a diasporic community relates to transnationalism as a process. Whether or not people of the same ethnic group scattered across the planet see themselves as belonging to a certain community is about the ongoing linkages and flows of ideas and values that lead them to imagine themselves in that way. For global diasporas, transnationalism corresponds to travel, the exchange of cultural ideas, the movement of objects and the sharing of cultural practices and customs. The concept is used in an attempt to appreciate how the 'global-ness' of political identities or movements is something that requires transnational practices to create them.

This links to a second way in which cultural geographers have made extensive use of the concept to describe the complex flows of cultural 'things' in today's world. The transnational geographies of cultural objects has been a particular focus of attention (Crang et al. 2003). As we discussed when considering geographies of consumption, many products and material objects are marked by the characteristics of particular places but travel great distances to be consumed in other places. Finally, transnationalism has also been used in relation to the emergence of a new 'transnational elite' social group or class in global society. Geographers have become interested in this new highly mobile segment of global society who may have multiple passports from different nation-states and who straddle both local laws and regulations, but also live across multiple cultural norms associated with different places. Essentially these are people who live across national boundaries in today's world, and are now, it is argued, important in the processes of cultural and political globalization.

TRAVEL AND TOURISM

One of the most important aspects of the rise in travel in the last century or so is the development of tourism. Tourism is now the world's largest industry, employing more than 240 million people (Adey 2009), and has an enormous impact on economies and societies around the globe as well as huge implications for the environment and the physical fabric of places. Human geographers have always been at the forefront of the study of tourism, probably because of the way in which debates about tourism combine economic, environmental, social and development issues. There is a large body of work on tourism within human geography, and it is impossible to summarize all of the many aspects to this here. For our purposes, however, we can consider three main aspects of geographical work.

First, a major area of work has considered the development of travel and tourist infrastructure, the planning process and the impact of such development on economies and places. From the 1970s onwards, human geographers mapped and measured the rise of tourism as an industry in terms of its economic impacts. This included examining where it was creating jobs, the nature of the companies that developed and the kinds of tourist developments that appeared in certain places. Economic geographical work has also looked at the role of tourism in economic development in lower-income regions, and considered the role of governments and other agencies in facilitating or hindering tourist development. However, a more substantial body of work in geography has appraised the wider impact of the rapid growth of tourism in the last 50 years, and developed a range of critical lines of analysis in relation to tourism. Examples include how environmental geographers have pointed to the negative impact of poorly regulated tourist development that has damaged natural habitats (for example, marine coral ecosystems) (Desforges 2005; Mowforth and Munt 2008). More recent work has considered how viable different **ecotourism** projects might be to develop more sustainable forms of tourism, as well as the rise of new forms of 'ethical tourism' (see box on voluntourism).

Second, work from a cultural geographical perspective has engaged with the social and cultural values that underpin much of the tourist industry (Williams 2009). Geographers have assessed the commonly held view that excessive tourist development 'destroys places' and 'spoils local cultures'. Increasingly, cultural geographical work has been important in questioning and problematizing these ideas. Human geographers point to the complex relationship between places and people's culture, and the difficulty in isolating 'pre-tourist' or 'authentic' local cultures. The debate about tourism has thus been important in developing theoretical understandings of how place 'authenticity' is constructed and how classic concepts of authenticity in the global South are in fact caught up in colonial discourses – authentic often means 'primitive', 'working-class', 'peasant'. In this sense, geographers have argued that the 'authentic places' tourists are offered and seek are presented as isolated from the 'outside influences' of other places when they are not isolated at all. In the globalized world we live in today, this represents a false idea of the world since communities everywhere on the planet are increasingly linked to each other by all kinds of flows and connections.

Finally, more recent work within human geography has moved beyond simple critiques of tourist development to develop a more sophisticated understanding of what happens to places when they become connected to the global tourism industry (Church and Coles 2006). Tourism in this sense produces places. Recent work has argued that it does this because of the distinctive relationship it sets up between producers (those who provide tourist infrastructure and services) and the consumers (tourists). Tourists often meet the producers of tourist experiences face to face – that is, they are in the same physical place. What is more, the experience of tourism cannot be taken home but involves the consumption of places by actually being there. If you visit the Empire State Building in New York or Machu Picchu (the lost Inca city) in Peru, what you consume is the experience of being there. You look at or 'gaze' on the place, but may also experience it in other ways – food, physical activity and so on. Geographers have thus become interested in the range of actors that influence the nature of these tourist places, whether they are travel companies shaping the nature of hotels and facilities or tour guides leading tourists around a certain set of imagined spaces. Such work also extends to a consideration of how the development of tourist places as destinations shapes patterns of global travel and the way in which tourist spaces are regulated.

'VOLUNTOURISM', GAP YEARS AND THE GLOBAL SOUTH

Many young people (and some older people), mostly from countries in the global North, have become involved in a new phenomenon in today's world: the emergence of volunteer tourism – sometimes termed 'voluntourism' – in places of the global South. If you live in the UK or Australia, this might correspond to some kind of 'gap year', 'career break' (for people of working age) or a post-retirement expedition. In the US, it is more likely that younger people will experience voluntourism through organizations such as the Peace Corps. Its common feature is individuals spending time on a volunteering project – that is social, educational or environmental in nature – in a country in the global South.

Work within human geography has been well positioned to try to understand the complex factors behind the rise of 'voluntourism', 'gap years', international voluntary service and other related phenomena in our globalized world. Media critics in Western countries have, over the last decade or so, sometimes dismissed many types of voluntourism as just another form of (often backpacker) tourism with little benefit beyond that of other forms of tourism. However, geographical thinking has argued that this kind of phenomenon is much more complicated. It is neither a pure form of tourism nor work (see section on work below), and is linked to the internationalization of the labour market, the need for people increasingly to gain experience of different regions and the transformations in the nature of what tourism itself means. For example, the rise of environmental awareness and discourses of environmental action and conservation are clearly linked to the rise of ecotourism and associated forms of voluntourism. It is also impossible to appreciate effectively the factors behind the development of this form of mobility without an understanding of the blend of economic, cultural, social and technological factors driving it.

LABOUR GEOGRAPHIES

Human geography's interest in labour extends beyond purely economic concerns about the nature of labour and its role in producing goods and services in places. During the 1970s and 1980s, a significant body of work within a political economy tradition in human geography also examined labour inequalities and the politics of labour. Human geographers were at this point interested in the factors producing labour exploitation in different industries and places, as well as the way in which labour organizations such as trade unions affected working conditions, workers' rights and industrial relations. Since the 1980s, this strand of work has continued but has become increasingly concerned with labour inequality at the global rather than the national and regional levels. Geographers today are concerned, for example, with the nature of working conditions that workers in textiles or electronics factories in Asia experience and how that relates to union organization or the role of non-governmental organizations and global consumers pressurizing firms to improve them.

WORK

Beneath the scale of the national or regional labour market, or the labour force in a certain industry, human geography has also been very concerned with individual labour – what is more generally known as 'work'. Understanding what work 'is' as a form of activity people undertake seems at face value to be straightforward. The commonest everyday conception of work is something people do for money (i.e. they are paid). However, human geographical thinking about work joins other social science disciplines in conceptualizing this activity in a rather more complex and diverse manner. Work can be paid or unpaid, the latter category including, for example, voluntary work or domestic work (cleaning your house or doing the laundry). Furthermore, there is then the question of the spaces in which work occurs. Conventionally, we think of people working in factories or offices but in fact work occurs in all kinds of different environments. Domestic work obviously occurs in the home, but people obviously also work in all kinds of different places: in fields, in the street, online, while they are travelling on trains, ships and planes (whether as passengers or crew). In the 'new economy', where services have been ever more dominant in terms of employment, a growing proportion of people also undertake paid work in service industry environments of one form or another: in shops, malls, restaurants, bars, hotels, gyms, cinemas and other leisure facilities. Work is therefore a very broad concept, but human geographers see it as very geographical in nature because work remains deeply place-based. People when they are working, in short, have to be *somewhere* and in that sense work always occurs in a specific physical location (it also even occurs in virtual spaces). It may be a factory, an office, in a field or in the street, but when someone undertakes work, they are clearly situated in one place or another (Castree et al. 2004).

THE GLOBALIZATION OF WORK AND THE EMERGENCE OF A GLOBAL BUSINESS CLASS

Recent geographical thinking has begun to consider the consequences of ongoing globalization processes for work along with the way in which this is producing new kinds of worker identities. There are at least three important aspects to geographical thinking worth highlighting here.

First, geographers have begun to be interested in the globalization of working practices. The growth and increasing dominance of transnational firms in all industries in the global economy has led to a shift in the way in which people work. For many professional and managerial workers, there has been a significant increase in the need for business travel in order to conduct the business of transnational firms operating in many regions. For example, if you design computer games for Sony in the UK to be sold globally, it is likely you will need to visit colleagues in the process of designing a new game. Equally, if you are corporate lawyer in Hong Kong working on contracts to do with new factories for a Korean firm, you will almost certainly end up in meetings in Seoul. The point is that as the economic activity and firms transnationalize, working practices for a significant proportion of the world's workers are also globalizing.

Second, geographers have become interested the way in which work as an activity is influenced by factors across many scales. People may always and everywhere undertake work in a specific place, but what they are doing as work is increasingly linked to very distant relationships. Work as a kind of practice is increasingly becoming globalized in places just as workers are becoming globally mobile. There are many aspects to this but improving information and communications technologies (ICT) are key factors in facilitating this change. Many transnational firms make extensive use of tele- and video-conferencing, Skype and other ever-improving forms of virtual communication enabling work to be conducted by workers together while they are physically scattered across the globe. However, it is not just those working for large global firms that are undertaking forms of globalized work. The work undertaken by agricultural workers in Kenyan flower- or vegetable-growing industries, for example, is closely regulated and shaped by the needs and demands on a daily basis of supermarkets in Europe and North America.

Finally, arguments related to the globalization of work have also been applied to the question of what is happening to worker identities and groupings in the global economy. Traditionally, human geographers, like other social scientists, made use of the idea of class within regions or national economies to differentiate groups of workers by their common kinds of work and the relationship of this to the overall production process. The working class undertook manual and other forms of labour while white-collar 'middle-class' workers occupied professional and managerial positions. While the concept of class has been widely criticized and it is argued in many countries that these kinds of class distinctions are no longer very clear cut, at the global scale, the concept of new forms of emerging class groups has been proposed. In the section on transnationalism, we mentioned how geographers have become interested in the emergence of a transnational business elite in recent decades. Some geographers and other social scientists think that, in relation to economic processes, a segment of this elite really corresponds to a new kind of global class (see box). This class is composed of people who occupy managerial and professional positions in that economy, generally working for the growing number of transnational corporations in that economy. Often these are the highly mobile business travellers who undertake the globalized working practices already discussed. These people come from and work in many different countries, and in that respect quite often have more in common with each other than in that economy with many of the people from their home country.

Whether or not elite workers in the global economy do in fact correspond to a new kind of global class remains the subject of much debate, but human geographers concerned with labour and work are continuing to study the emergence and development of this new layer in global society.

THE NEW GLOBAL BUSINESS NOMADS?

In the media, global business executives are represented by positive images of sophisticated travellers who enjoy the excitement, luxury and cosmopolitanism of travel often seen in James Bond films. However, the reality of transnational business work for business people in the global economy has become less romanticized. George Clooney in the 2009 film *Up in the Air* depicts a more contemporary representation of the experience of what some human geographers call the 'new global business class'. These are the executives who work for the growing number of transnational corporations in the global economy, and who have to spend an increasing proportion of their time engaged in 'nomadic' business travel in order to run these companies. Clooney's character falls in love with a fellow traveller but their relationship nearly fails because of the demands of such a mobile lifestyle. Human geographers have pointed to the limitations of global mobility and the negative impacts on workers in terms of family life, relationships and happiness. Clearly there are limits to the degree to which workers can be physically mobile, and the reality of increased business travel for many workers is a detrimental impact on their quality of life far removed from the glamorous images.

SUMMARY

In this chapter we have considered:

- The way population geographers have studied global population unevenness, change and crisis, as well as theories of how population growth relates to economic development;
- The concept of citizenship and how it varies geographically, as well as the relationship between citizenship and different types of spaces;
- The nature of diasporic communities;
- Geographical approaches to mobility, travel and tourism;
- The way in which geographers have theorized different types and patterns of migration;
- What is meant by labour geographies and how human geographers have studied difference in labour markets and labour relations between different regions of the world;
- The nature of work and how work has become globalized as an activity;
- Debates about the emergence of a global business class in recent decades.

FURTHER READING

Adey, P. (2008) *Mobility*. London: Routledge.
One of the Key Ideas in Geography series, this books provides an overview of the breadth of current human geographical debates about mobility.

Castree, N., Coe, N., Ward, K. and Samers, M. (2004) *Spaces of Work*. London: Sage.
This book has a good discussion of labour geographies and how labour in the global economy is caught up in place. Chapters 4, 5 and 7 are especially useful in this respect.

Cohen, R. (2008) *Global Diasporas: An Introduction*. London; Routledge.
The author is not a human geographer, but on the question of diasporas this book is probably one of the best overviews of the issues that concern geographical work.

Pacione, M. (ed.) (2012) *Population Geography: Progress and Prospect*. London: Routledge.
This is a classic geographical population book from the mid-1980s that has been republished. It has a good range of chapters discussing different ways in which human geographers have engaged with questions of population.

Williams, S. (2009) *Tourism Geography: A New Synthesis* [2nd edition]. London: Routledge.
This book in its new edition provides a good overview of the range of issues addressed in relation to tourism by human geographers.

WEB RESOURCES

Have a look at the UN Population division's website: www.un.org/esa/population/
On transnationalism, the University of Toronto has an interesting research centre www.utoronto.ca/cdts/graduate.html

BODIES, PRACTICES AND IDENTITIES

This chapter examines how human geographers have engaged in debates about the nature of human bodies, social practices and the development of senses of identity surrounding gender, race, sexuality and age. All three areas of geographical work on these topics span debates in many of the major sub-disciplines within the subject including social, cultural, economic and political geographies.

THE BODY

It may seem a little strange that human geographers are interested in human bodies, but in fact a substantial amount of geographical work has focused on the nature of bodies within social, cultural and feminist geographies. What is a body? Each of us has one but there is actually much debate among human geographers and other social scientists as to where we might think about the body 'beginning' and 'ending' and about what it means to have a body.

Taking the first issue, geographers' arguments are based on the relationship between the mind and the body, and whether or not the body is a 'natural' or 'social' thing (Valentine 2001). Geographical thinking has engaged with the widespread conception of the relationship between the body and mind first developed by the 17th-century philosopher René Descartes (1596–1650). Descartes is famous for the widely (and often incorrectly) quoted phrase: 'I think, therefore I am.' The notion (often termed as 'Cartesian') expresses the idea that your body is essentially like a machine, with your mind as the controlling power that is responsible for your intelligence, your identity and your spirituality. This is what is known as a dualistic division and has had many implications for how we understand the world centuries later.

Feminist geographers argue that this dualism is important because it has shaped understandings of society and the nature of space itself, as well as the way geographical knowledge has been produced (Rose 1993). For example, the classic definition Descartes offers of the mind as being separate from the body is argued to have led to a definition of rational knowledge as masculinist. This means that any knower of knowledge can separate themself from their body. 'Emotions, values, and past experiences' are conversely feminine, and have no place in rational or scientific knowledge (Rose 1993: 7). In this way, knowledge is not tainted by bodily experience and thought is context-free and autonomous.

Another related way in which human geographers have theorized what it means to have a body concerns the degree to which bodies are 'natural' or 'social'. Echoing the debates we discussed earlier concerning the degree to which it is possible to separate the category of 'nature' from 'society', human geographers have argued that many of our understandings of bodies as 'natural' are in fact socially constructed.

BODY AS SPACES

In geographical thinking about the body, the key issue is that the body does not just exist 'in space', but is actually a space in itself. Human geographers have conceptualized the body as a space in three ways. First, they argue that it is a kind of surface that is marked, inscribed and transformed by culture. We inscribe our identities on our bodies, and it is also a surface that is written on, marked, scarred and transformed by wider society and its institutions. Geographers share with other social scientists an understanding of how the development of modernity affected the way in which people think about their bodies and behave. In Shakespearian times, for example, social attitudes to spitting, defecating and personal hygiene were very different – especially in public places. People were not offended by human excrement in the street. It was not just because there were no public toilets; society saw no need for them. The point is that attitudes towards bodies and their behaviours are not pre-given but have evolved over time and our bodies reflect and are governed by societies' expectations, norms and rules. Equally, the body is a surface space as it reflects social position. People in a certain social class wear certain clothes, behave in certain ways and eat certain foods.

A second way in which geographers have understood the body as a space is as the space through which we sense and experience the world. A body is a space that senses (i.e. it is 'sensuous' in the literal meaning of the word). The body is therefore a *personal space* through which people come to construct widely held definitions of ideas such as well-being, illness, happiness and health. Bodies are also the means by which we can experience and connect with other spaces. This has led human geographers to argue that there is a need to appreciate the nature of knowledge from all our senses, not just the visual (Thrift 1996). The experience of a disease such as myalgic encephalomyelitis (known as ME) may mean that a person who looks young, healthy and athletic and mobile is in fact afflicted by fatigue, immobility and discomfort (cf. Moss 1999, cited in Valentine 2001).

Finally, a third geographical understanding of the body has focused on the way in which our bodies are the main location where our personal identities are constructed. Who you are and how you see yourself as a person is very closely bound up with the nature of your body – whether you are a man or a woman, the colour of your skin, your height, size and appearance. Human geographers have also drawn on psychoanalytical theory to think about how we understand the difference between ourselves (our 'interior selves') and the external world, arguing that interior ideas of self lead to projections in wider society and across space of 'us' (self) and 'them' (other).

BODIES AS SPATIALIZED PROJECTS

One particular way in which human geographers have understood bodies as surfaces warrants further attention. This is the way geographers have theorized bodies as surfaces that act as symbols of the self (and also see the section on self and other below). In Western societies bodies have become a spatialized 'project' to be worked on. Cultural geographers relate this emergence of the body as a project to a range of factors including increasing mass consumption, the democratization of culture, a decline in religious morality and a post-industrial emphasis in Western societies on **hedonism** and pleasure (Turner 1992, cited in Valentine 2001). People thus construct identities through their appearance, and the media, advertising, fashion, medicine and consumer cultures all shape discourses within which we evaluate and understand our bodies.

Consider the ways in which plastic surgery and sports science have given us the possibility to control and reconstruct how our bodies appear. Not everyone wants to do this – or can afford to do so – but the body has become an adaptable thing in terms of its size, shape, appearance and so on. The body has become linked in Western cultures to a project of identity construction. Women aspire to being a size eight, to having a well-proportioned figure or skin that looks a certain way. Society shapes this set of expectations in terms of health campaigns that encourage people to, for example, lose weight, take exercise or avoid an excessive amount of sunlight (if you live in a country such as Australia or the US). People also seek to counter the effects of ageing, with being old implicitly constructed as an undesirable form of identity. Likewise, overweight people are framed in discourses that often stereotype them as indulgent or lazy. Identity construction is also reflected in more explicit bodily practices such as body-building or tattoos and piercings.

Human geographers argue, however, that the process of working on the body as a project is always partial and incomplete. Bodies are not completely controllable and people experience conflicting ideas of how their bodies should be produced as a space in different places and at different times. People experience contradictory impulses all the time – for example, feeling pressure from society to diet in order to appear slim in public while still enjoying high-calorie foods in the private spaces of their homes or when dining with others.

BODIES IN SPACE

There are two elements to the way in which geographers theorize how bodies exist in space. The first is the way bodies 'take up' space insofar as they 'occupy' it through size and appearance, as well as by how they move and are moved in space (termed 'comportment'). With respect to the former, social geographers have been interested in, for example, how the design of many environments – airplanes cars, buses, restaurants – fail to accomodate fat people and make them uncomfortable as well as marking them out in public as 'oversized'. In terms of the latter, feminist geographers in particular have pointed to the way women occupy and use space differently from men. They often demonstrate restricted bodily movements such as sitting with their legs crossed and their arms across themselves (Valentine 2001), whereas men are more likely to sit with their legs open and using their hands to gesture. This is not because women are weaker than men, but because they approach tasks differently. Women think they are weaker and act accordingly, experiencing their bodies as more fragile than they are (Young 1990). They may also be fearful that their body space may be invaded by men, and therefore experience their bodies as 'enclosed' or separate from public space. Comportment in this sense refers to the physicality of bodies, understood through a range of social meanings and discourses.

The second aspect of the concern of human geographers for bodies in space centres around how our material bodies are the basis of our experience of everyday spaces. Everyone else reacts to *our* bodies and we react to them and read stories off about a person (such as age, lifestyle, politics, identity, etc.). A key area of work in geography has thus focused on how the nature of our bodies makes a difference to how we experience places and how different bodies exist in space in a way that marginalizes or excludes them. If you are a young person you may experience somewhere very differently from an older person. If you are a man, certain places will seem very different than they would if you were a woman. Think about the experience you would have as a young man walking around a large city at night as opposed to a young woman or as a person in a wheelchair trying to navigate a subway system in a city such as New York or Tokyo.

GENDERED BODIES IN THE PROFESSIONAL WORKPLACE

In his 1989 semi-autobiographical novel *Liar's Poker,* Michael Lewis describes the heavily gendered culture of work as a bond trader in Wall Street during the boom times of the 1980s. The most successful traders were described as 'big swinging dicks', in an overt equivalence to male sexual prowess. Lewis describes a male-dominated world where not just the language but also the behaviours and work cultural norms are strongly masculine. Similarly, Tom Wolfe's novel *The Bonfire of the Vanities* (1987) portrayed these traders as arrogant masculine 'masters of the universe' ruling the world economy.

In light of the cultural turn, geographical work investigated the nature of these heavily gendered financial service and banking workplaces. In her 1994 book *Capital Culture*, Linda McDowell argued that gender relations within banking and other financial firms were caught up in a complex set of discourses, embodied practices and power relations. The 1990s had seen a growing proportion of women employed in banking and other leading financial service sectors, but they were concentrated in the lower-grade jobs in these organizations. Very few women ever became directors or senior managers.

Based on research in the City of London, McDowell showed how the embodied gender performances (see section on performance and performativity below) of men and women related in complex ways to the way work was undertaken in these banks. An important finding was that women did not rise to senior positions not because they were explicitly discriminated against, but because they were less convincing in the necessary 'gendered performances' expected of senior managers in banks and other firms. McDowell thus showed how these firms exhibited gendered cultures of work that cannot be easily changed by simply altering policies within firms or legal regulation.

PRACTICES

A broad definition of practices as used by human geographers and other social scientists corresponds to 'the actions of individual or groups'. This conceptualization of action includes not just physical behaviour but mental activities such as theorizing or learning. Yet like many such generalized concepts, practice has a more specific and distinct meaning within a number of schools of social scientific thinking. In the wake of the cultural turn, human geographers have consulted a range of work in sociology, anthropology, psychology and management studies in order to think about the significance of the spatiality of practices and of their consequences in a number of ways. At least three different strands of thinking about practice have been particularly influential (Jones and Murphy 2010).

The first concerns how practices help structure, organize and govern cultures, societies and nations. Second, human geographers across a range of sub-disciplinary areas have been interested in what we might call 'communication' and 'discursive' practices. These include the role of all forms of social performance, social communication and language in shaping societies, economies and cultures. Of particular relevance to this is the social science literature concerned with what is known as **actor-network theory** (ANT) (Law 1993; Latour 2005), which is being increasingly used by human geographers. ANT argues that all forms of communication practice offer insights into the ways and means of what is termed *translation* – the process through which actors exert power, mobilize material objects, and 'perform' in order to achieve particular objectives.

Finally, human geography has made use of sociological work that addresses how practices embody tacit forms of knowledge and how they contribute to organizational cohesion and collective learning. For example, the managerial and knowledge creation practices relied on in particular industries and transnational firms (Amin and Cohendet 2004; Glückler 2005), the governing practices of elites and states seeking to control and direct economies (Larner 2005), and the alternative or 'ordinary' practices that are involved in 'non-capitalist' economic forms such as cooperatives, informal livelihood strategies, or unpaid labour (Lee 2006; Gibson-Graham 2008).

PERFORMANCE AND PERFORMATIVITY

These concepts have a slightly different use in human geography, with three meanings that are used in the subject. The first is the everyday one involving such activities as music, dance and acting. Human geographers have begun to research these topics in relation to the way in which subjectivity and identities are constructed and linked to spatial experience as well as the visual world. Second, there is a substantial literature that has utilized sociological understandings of social action as being like 'scripted performances', framed by a range of discourse-based understandings of the social world. This draws on the philosopher Judith Butler's (1990; 1993) rejection of the idea that biology is the foundation for the categories of either sex or gender. As mentioned above, gender is not what you are, it is what you do (Pratt 2005) and is therefore part of the process of an individual subjectivity. Understood as this kind of scripted performance, gender corresponds to 'a repeated stylization of the body' and is a 'set of repeated acts with a highly rigid regulatory framework' (Pratt 2005). The idea is that the regulatory framework of gender categories has 'congealed' over (very) long periods of time in human history, which have led to them *appearing* to both have substance and be natural ways for people to exist. Human geographers have used this idea of gender as performance to understand the nature of workplace practices as well as geographical context in the formation of sexual identities.

A third way in which performativity is used within the subject relates to work within sociology, anthropology and socioeconomics which has become increasingly concerned to understand how economies are 'performed' (Callon 1998; Mackenzie 2008). This work beyond geography argues, for example, that what happens in global financial markets cannot be explained successfully through the market theories of neoclassical economists, but rather as performed sets of practices of various actors (traders, investors, etc.) making use of material environments, devices (trading floors, computers, telecommunications) and conceptual tools such as mathematical models. The point is that a combination of interaction with the material world, cultural values, emotions and technical devices contributes to producing the eventual outcomes.

EMOTION AND AFFECT

Emotions shape the nature of many aspects of social life, and therefore have geographical impacts. The term 'emotional geography' is used to describe how spatial knowledge of the world is written with or on emotions, whereas a geography of emotions relates to the mapping of different emotions and emotional states (Parr 2005). The social world is 'constructed and lived through human emotions', such as 'pain, bereavement, fear, elation, anger, love and so on … ' (Anderson and Smith 2001). In general, human geographers understand emotions as things people acquire through culture that are bound up with conscious actions but also with the experience of living. Social, cultural and feminist geographers have been concerned with a wide range of different emotions and their impacts that relate to the social and material world – emotions impact not only on everyday life but on politics, warfare, landscapes, the home, the built environment, etc.

Related is the concept of 'affect'. This is an abstract and difficult idea that refers to the way in which any representation fails to capture completely the way in which social life is full of emotional and constantly changed 'lived experience'. Affect is therefore an experiential phenomenon that attempts to understand how bodies are altered by emotions ('affected') and also have emotional impacts on (that is, 'affect') others. The important point is that this is not just related to individuals but is a collective phenomenon in society. Affect is therefore not a conscious or intended consequence of social action but is just an unavoidable aspect of being alive. Human geographers have sought to map 'affect' (Pile 2010), using both it and ideas of emotion to understand, for example, the nature of bodily movements in space and their relationship to emotions. Work has looked at aspects of social life including dance, theatrical performance and game-playing (Thrift 1997) (see box).

UNDERSTANDING 'AFFECT': THE CASE OF COMPUTER GAMES

Most people have experienced playing a computer game that, in the 21st century, can immerse them in a complex alternative world. In the last decade or more, of course, online games have allowed them to play in virtual worlds with millions of other gamers on the internet, often with an 'avatar'. Examples include games such as *World of Warcraft*, *Grand Theft Auto* or *Second Life*. These game worlds have a virtual geography and spatiality, and illustrate well the relationship between understanding such experiences as representations and the usefulness of the concept of affect.

In this respect, Ian Shaw and Barney Warf (2009) argue that we can understand video games as a kind of sensory commodity that exposes players to a variety of affects – largely emotional feelings of surprise, fear, anger, disgust, sadness and joy. They also arouse the body: just watch someone playing and the expression on their face as well as their behaviour. Shaw and Warf use this analysis to show how capitalism (computer games are commodities you pay for) has an emotional aspect that operates at the pre-cognitive (i.e. unconscious) level as well as at a conscious level. The concept of affect thus helps them understand how playing has impacts in the world beyond the computer screens. Game designers now focus on the emotional response to games, and the interactive embodied experience of playing a game involving shooting Arab (or 'other') enemies has wider impacts on the social world (see section below on self and other). Shaw and Warf (2009) argue that these game spaces are increasingly 'affective landscapes' where, as a player turns their attention to the experience of the space they are shaped not just by the representations of those spaces but also by the body's affective articulation in another world. Computer games are experiential and lived, and have to be understood as such, not just as representations.

IDENTITY

As the Introduction outlined, human geography today takes what is known as a 'non-essentialist' approach to identity, theorizing our identities as relational insofar as they are constructed in relation to how we see similarities and differences in other people. This relational perspective thus argues that no identity is 'innate' – that is it does not exist automatically. This represents a rejection by human geography today of earlier ideas about identity that saw them as being quite 'fixed' – you were a black person, a lesbian or middle class. Instead geographers now view identity as being in a constant process of construction and that people actually have *multiple* senses of identity. You can take on the sense of being a national citizen, a woman, a student, a tourist or any number of identities, and human geographers are interested in how these are shaped both by imaginative geographies, and the spatial context that people exist in. For example, if you live in a European country, you may not have a particularly strong sense of any Europeanness. However, if you take a holiday in China or Japan, for a whole range of reasons you may feel more European and identify with that identity. Related to this, geographers have been particularly interested in wider social science debates about the 'hybrid' nature of identities that argue that all identities – like all forms of culture – are hybrid mixtures that are in a constant state of change.

SELF AND OTHER

The self is a concept that seeks to express the way that senses of oneself are inevitably caught up in a relational process of definition based on differences between the self and someone or something else. The 'other' therefore refers in the abstract to a person or thing that is opposite or different to oneself. In that sense, the description of 'otherness' refers to the qualities that the other possesses that are different from those of the self.

Human geography over the last two decades has become ever more interested in taking the other seriously by thinking about 'different kinds of people who are situated in different kinds of spaces' and places (Philo 1997, cited in Cloke 2005b). Key to this is understanding how these others experience these spaces and places. This breaks down what can be termed '"the arrogance of the self", that essentially corresponds to the assumption that others must see the world the same way we do' (Cloke 2005b: 69). It has been argued that this amounts to being 'locked in the thought-prison of the "the same"' (Philo 1997) – meaning essentially that it is often almost impossible to appreciate the world from another person's perspective. The temptation is either to try to incorporate others into our sameness, or to exclude them. Either activity is, of course, a highly political action.

Geographical work has thus focused on producing knowledge of a whole range of people who have been 'othered' and whose experiences and worldview remain hidden from the supposedly objective social science that characterized human geography prior to the cultural turn. The other's experience is different across space and between places, as well as being shaped by certain geographical way of imagining the world. Classic examples include people who live in rural areas (Philo 1992) but also those excluded on grounds of class, disability, age, race, gender, sexuality and so on.

GEOGRAPHIES OF GENDER

Work on gender in human geography began in the 1970s and 1980s with feminist geographical work that sought to expose the disadvantaged position of women in society. Studies analysed, for example, the exclusion of women from labour markets in different places (Perrons 1995). It was also argued that, up until the 1980s, human geography had largely ignored half of the social world by excluding almost everything seen to be feminine (Monk and Hanson 1982). They offered plenty of examples: childcare, women's labour in the home, voluntary work, caring for the elderly, subsistence ways of making a living.

This kind of work has continued, but since the cultural turn the scope of geographical work on gender has become much broader. Feminist human geographers made extensive use of the feminist argument that 'rational knowledge is actually grounded in the position of white, heterosexual men who tend to see other people not like themselves only in terms of themselves' (Rose 1993). Geographers have sought to understand how women have been excluded from – and also, perhaps most importantly, have transformed what were understood to be – legitimate or worthwhile topics of geographical knowledge beyond those earlier 'feminine' topics. Until the last couple of decades, the questions of sexuality we consider below were regarded as inappropriate topics for human geographical study (Longhurst 1997). Moreover, there is now a large amount of work in human geography examining the geographical nature of gender differences in a variety of contexts. During the 1990s, for example, social and cultural geographers considered the gendered nature of landscape representation and cartography (Rose 1993).

Questions of gender were also applied to traditional questions dealt with by economic geographers. Beyond the scale of individual bodies already considered (McDowell 1997), another good example is the work of Gibson-Graham (1996), which controversially argues that the kinds of classic representations of globalization discussed in Chapter 2 were organized through masculine representations of capitalism and a metaphor of penetration and rape of feminized, vulnerably local economies (cited in Pratt 2005). The goal of Gibson-Graham's theorization of capitalism in this way is to break down masculine discourses that give power to key actors in the global economy (such as transnational corporations) by representing them as masculine bodies, but it remains debatable whether it actually succeeds in liberating alternative imaginings of the capitalist world.

GEOGRAPHIES OF SEXUALITY

Human geographical work on sexuality began in the 1970s by trying to understand the development of 'gay neighbourhoods' in the cities of the global North (Binnie and Valentine 1999). In this work, geographers examined the factors that led to the clustering of gay businesses (bars, clubs, shops) and drew gay men to these areas (Winchester and White 1988). This led to an increasing interest in the relationship between these gay spaces and capitalism in terms of urban land markets and the deliberate marketing of certain areas as 'sexualized spaces' (urban 'gay districts') (Binnie and Skeggs 2004).

However, since the cultural turn, geographical analysis of sexuality has broadened considerably to consider how all forms of sexual identities are shaped by social relations across space. It has also become increasingly interested in the tendency of popular culture to present heterosexuality as normal – termed 'heteronormativity'. In this second strand of work, human geographers have undertaken a great deal of analysis concerning what it means to be heterosexual (straight), bisexual, lesbian or gay and how this varies across time and space (and therefore between places). It is also often argued that human geography in effect took a 'queer turn' as this body of work mushroomed in the 1990s (Browne 2006). Good examples of this kind of work include Gill Valentine's study of lesbian experiences in the urban landscape (Valentine 1993) or Kath Browne's more recent work on lesbian and gay spaces of leisure such as music festivals (Browne et al. 2009).

A third aspect to geographical thinking about sexuality has been to begin to broaden out the scope of analysing othered sexual identities (lesbian, gay, bisexual) to think about the 'same' – heterosexuality. Social geographers became interested in marginal forms of heterosexuality. This includes a growing amount of work on prostitution (Hubbard 2004), which has examined, for example, how sex work is important in shaping the nature of districts and neighbourhoods in city spaces (see box). Other work has looked at sex tourism and its role in shaping the nature of certain tourist destinations (Jacobs 2010) in addition to research that examines the differences in the nature of heterosexuality in different places – for example, what it means to be heterosexual in a rural as opposed to an urban space (Little 2003). More recently, human geographers have become increasingly interested in what we can term 'moral' forms of sexuality and the impact of these discourses on politics, practices and the built environment.

Finally, geographies of sexuality over the last decade have broadened the analysis of sexuality to a whole string of wider issues addressed by social, cultural, political, urban and economic geographers. Geographers have begun to analyse, for example, the significance of non-heterosexual consumers in the global economy, the significance of lesbian and gay political networks and their emerging trans-nationality and the politics of homosexuality in global religious networks (Vanderbeck et al. 2011). Human geography is thus increasingly integrating the analysis of sexual identities into an analysis of every aspect of social life and its spatiality.

A TALE OF TWO CITIES: THE POLITICS OF PROSTITUTION IN LONDON AND PARIS

Many large cities across the planet have 'red light districts', where there is prostitution (sex work), and many of these districts have existed for centuries. Such areas are often regarded as undesirable features of cities, associated with illegal activity, exploitation and a threat to public order. But in the 21st century, there are dynamic politics around sex work, which have become very evident in large capital cities such as London and Paris. In both cities over the last decade, authorities have tried to clamp down and prevent prostitution in even the long-established areas – Soho in London and the Pigalle and Bois de Boulogne areas in Paris.

The urban geographer Phil Hubbard has examined the development of 'zero tolerance' policies by urban authorities to sex work (Hubbard 2003). City governments imposed new, steep fines and the threat of imprisonment in an attempt to 'clean up' these red light districts. Hubbard explores the contested politics of public space in these cities that has emerged, exploring the motives for trying to exclude sex workers and the political reaction and protests this has produced from the sex workers themselves. In Britain and France, sex workers have formed unions and mounted protests against the increasing criminalization of prostitution. Hubbard argues that the new 'zero tolerance' strategies are closely related to capitalist strategies to market districts such as Soho in London around a family-friendly conception of sexualized space. In this way, he shows how moral discourses about sex are caught up in the nature of public spaces, in urban politics and in the urban economy.

GEOGRAPHIES OF RACE

As a category, race is an idea that distinguishes between people on the basis of physical differences secondary to their bodies. The most obvious is skin colour, but other characteristics are also associated with it (a person's hair or physical characterisics). Historically, race was of course linked to notions of the racial superiority of 'white' Europeans through the period of colonial expansion from the 16th to the 19th centuries. Although any biological basis to any strict division of human beings into different races has long since been discredited, the category of course persists as a political and cultural construct. In particular, it forms the basis for 'racism', which is a political ideology of difference between individuals.

Within human geography, work on race has to a large extent mirrored that on those 'othered' identities we have been discussing insofar as it began with a consideration of racial identities which are not the 'same' (in this case 'white'). At least three main strands of work on the geographies of race are worth identifying. The first of these is similar to that on sexual identity insofar as human geographers have been concerned to map, analyse and understand the nature of racial segregation within certain spaces: most notably, cities, regions and nations. In a straightforward way, human geographers have mapped how certain racially categorized social groups are concentrated in specific areas. In this respect, there is a considerable amount of work on 'ghettoization' in cities of the global North where ethnic minority groups are concentrated in areas of high unemployment, urban deprivation and poor state services (Nayak 2003). Human geographers have also been interested in how certain ethnic groups become trapped in 'racialized' spaces that prevent them from taking up new opportunities elsewhere when they arise. For example, many of the poor black communities that were worst affected by Hurricane Katrina in New Orleans in 2005 were unable to move to new lives and employment opportunities elsewhere in the US (such as California).

A second strand of work is more concerned with the relationship between different forms of geographical knowledge and imaginaries about places or environments and ideas about race. For example, Sibley (1999) argues that way in which 'whiteness' is equated with purity, order and cleanliness in European cultures has produced the negative stereotypes that exists of Africans, Afro-Caribbeans, Indians and Roma gypsies. The argument is that each has been constructed as a marginal threat within various nationalisms, which itself has fuelled spatial segregation. Finally, a third aspect of human geographers' interest in race concerns the different experience of people of racial identity in different spaces and places. This has examined, for example, how being 'black' or 'white' in certain urban areas leads to fear and anxiety. White people may feel threatened in the 'black inner city', but conversely black people often experience similar feelings of anxiety in 'white' rural spaces in northern Europe or North America. A growing amount of more recent work has also been particularly concerned with the experience of 'whiteness' in particular times, places and spaces (Bonnett 2008)

CHILDREN'S GEOGRAPHIES

Human geography during the 1990s saw a rapid growth in interest in age, and in particular in the experience of children. The reason was largely a realization that the subject had ignored the very different lived experiences of children and young people in terms of space and place. Again, human geographers see the identity of being a child as something that is socially constructed, not some pure biological category (James et al. 1998). The qualities of what it means to be a child have therefore varied over time and between places, and human geographers have been concerned with the 'place-specific' nature of being a child, exposing many of the everyday assumptions people hold about childhood as being particular to the West and Europe. Examples would be the way children are less able and competent than adults and therefore need to be educated how to become adults, or how young children are 'innocent' and free from the responsibilities that adults have. These are culturally specific assumptions and do not really hold for the way children's identities are constructed in many countries in the global South. In fact, in the global South, many children make crucial and important economic and social contributions to family livelihoods through domestic, agricultural and all kinds of paid work (Punch 2001).

Human geographers have also examined in some depth the kinds of everyday spaces around which children's identities are constructed. They argue that all kinds of spaces continually produce and reproduce ideas about what it means to be a child. Think about the way ideas of what childhood means are expressed in the design and use made of the built environment. This of course applies to all kinds of spaces: homes, schools, leisure areas. Another dimension to this is the relationship of children to public space, with children seen as being both vulnerable in such spaces and also a threat to adult control when being unruly (for example, teenagers riding skateboards).

SUMMARY

This chapter has:

- Considered how human geographers have theorized the body in terms of bodies as spaces, bodies occupying space and bodies as spatialized projects;
- Discussed the growing significance of concepts of practice, performance and performativity in human geography;
- Examined the way in which human geography has become increasingly interested in the emotions and made use of the concept of affect to overcome the limitations of textual, visual and linguistic representation to know the social world;
- Further explored the conception of identity, self and other used in human geography, in particular examining how various identities based on gender, race, sexuality and age are socially

constructed and have a range of impacts on geographical knowledge and social life.

FURTHER READING

Brown, K., Lim, J. and Browne, G. (2009) *Geographies of Sexualities: Theories, Practices and Political.* Aldershot: Ashgate.

This is a good and up-to-date collection of essays on different aspects of the analysis of sexuality in human geography today.

Dwyer, C. and Bressey, C. (eds) (2008) *New Geographies of Race and Racism.* Aldershot: Ashgate.

This is good collection of essays about the way race is being analysed in human geography, although the examples are mostly drawn from the UK and Ireland.

Holt, L. (ed.) (2011) *Children's Geographies: An International Perspective.* London: Routledge.

The Introduction to this book gives a good overview on the current state of human geographical work on children's geographies with very good chapters on youth identity and families in the collection.

Valentine, G. (2001) Chapter 2 on 'The Body' in *Social Geographies: Society and Space.* Harlow: Prentice Hall.

Although a few years old now, this book remains one of the most comprehensive texts on social geography with the section on the body being particularly useful.

WEB RESOURCES

Many of the debates that geographers make use of about gender as 'performance' are the subject of wider debates in social science and policy thought. This website provides an introduction to these discussions: www.genderforum.org/home/

An interesting research centre at the University of Natal concerned with race and identity: http://ccrri.ukzn.ac.za/

8

CONCLUDING OVERVIEW: HUMAN GEOGRAPHY TODAY

This book began by arguing that human geography is perhaps unique among the social sciences in its breadth and scope. It is also distinctive in being a 'half discipline', tied closely to the natural science focus of physical geography. Yet, as pointed out in the Introduction, geography as a discipline (and human geography as part of it) has sometimes been criticized for being too diverse. One of the central aims of this book has been to dispel this idea. The numerous sub-disciplinary areas in human geography (cultural, political economic, urban geography and so on) do sometimes give this impression. Likewise the enormous number of topics that human geographers research undeniably reveals enormous diversity, but what the preceding chapters have sought to do is demonstrate how this diversity does not equate to incoherence. This book has provided what is undoubtedly a rollercoaster ride through a wide range of topics that also almost all concern other social science disciplines in one way or another. Yet what holds them together is what we have called a geographical imagination: a focus on the spatiality of social life and on the way social phenomena relate to each other across space. In this respect, you should end this book with a good understanding of how the various sub-disciplinary areas in human geography are actually very closely entwined together. The way 'economic' geographers theorize the nature of firms in the global economy today is increasingly caught up in the theoretical arguments made by social and cultural geographers regarding practice, performance, identity and emotion. Equally, the traditional concerns of 'development geographers' around places in the global South have become blurred with many of the concerns found in other subdisciplinary areas of the subject: for example, the significance of innovation and creativity in economic development or the emergence of transnational political networks.

Human geography is, therefore, a discipline whose power and significance lies in the kind of interdisciplinarity that is so often called for in other subjects. Economists, it is sometimes argued, need to appreciate matters from a sociological perspective. Business and management theorists need to engage with the ideas of anthropologists. Human geography does this almost as matter of course, because of its underlying cross-disciplinary range of interests. In this respect, it is, I would maintain, one of the most exciting and innovative of social science subjects.

CURRENT RESEARCH THEMES IN HUMAN GEOGRAPHY

Here it is obviously impossible to provide an overview of the many areas of research being undertaken by human geographers at present. Instead what I want to do is conclude this book with an indication of some of the major themes that characterize research in human geography in the early 21st century. These will at least give you an idea of the kinds of research topics you would encounter if you were to go to an international conference on the subject tomorrow. Even this list of themes cannot claim to be comprehensive, but it does at least provide a flavour of some of the most significant research areas human geographers are engaged in and their enormous relevance to the wider world. They are not listed in any particular order of priority either as all of these themes relate to important challenges facing the world.

First, as should be obvious from the discussion in Chapters 2 and 3, human geographical research is currently concerned with the future uneven development of the global capitalist economy. Aside from examining the nature of firms, industries and regional economies, geographers are investigating the discourses that frame ideas of how the global economy should work in researching the ideology of neoliberalism. Research is also concerned to understand how the global financial system operates, whether it could lead to a global economic collapse, and what this means for social justice and equality across the planet. Equally, geographical research is interested in the operation of capitalism itself: how the economy innovates, the experience of workers in different regions and the complex relationship between consumption, culture and what products and services are created. At a smaller scale, however, human geography has enormous relevance and utility for anyone thinking about how to better manage regions, cities and localities. Much work in economic geography is very closely aligned to questions of what causes economic decline, and how economic success in a global capitalist economy can be generated in different places.

Related to this is the substantial research considering shifts in geopolitical power worldwide and the ongoing transformation of what was once called 'the developing world'. Human geographers have been researching the nature of development in Africa, Asia and Latin America for more than a century, but today this research is concerned more with new questions of increasingly wealthier and more globalized societies in the global South. Human geography has addressed in its research the extent to which China and India are becoming global superpowers, but also why many African countries have been left behind compared with their economic success in the last three decades. Equally important at the local scale are the transformations occurring to communities across the world as a consequence of wider globalization processes, and much geographical research today is examining how identities, cultural ideas, politics and social values are changing. Political geographers also continue to research the changing nature of international politics, including an understanding of political resistance, and by what means global society changes and develops.

Another major theme of geographical research concerns the global environment and the politics that surround this. Human geographers are at the forefront of research into the politics of climate change, examining, for example, what kinds of strategies and government policies are likely to succeed in managing its effects in different regions around the world. That includes research on the development of sustainable energy sources, sustainable cities, a low-carbon economy and sustainable agriculture and food production. Research in human geography is also concerned with many more different kinds of environmental challenges: resource management, the preservation of global biodiversity and pollution and waste management. Many of these research topics link to the research undertaken by physical geographers.

Fourth, much research in human geography remains firmly concerned with issues of social justice and inequality. Aside from research into how capitalism produces and reproduces social inequalities, human geographers continue to be concerned with social justice linked to those questions of identity (race, gender, sexuality, age, etc.) discussed in Chapter 7. Geographers research such questions at a range of scales from the level of cities to global politics. Part of this research is also concerned with more conceptual and theoretical questions about those issues of representation, subjectivity and the nature of human knowledge of the world discussed in this book. Such research is perhaps less immediately applicable at first sight to government policies but plays an important role in generating new ideas about the nature of social life. Such ideas and theories often have unknown and unintended positive consequences as they filter out into the wider society. Human geographers working in these areas are thus contributing to a wider body of research in the social sciences and humanities that contributes to an ongoing shift in the cultural and social values that people share across the globe. Research that, for example, exposes the Western-centric nature of knowledge helps shift our understanding of cultures themselves and produces new kinds of political agendas.

Finally, a variety of research across human geography is concerned with the dramatic impact of new forms of information and communication technologies. In the 21st century, the pace of change is unprecedented as mobile web devices, social networking and geolocational technologies transform the way we live. Think of the impact of Facebook or Twitter – or of GPS technologies such as satnavs – on all aspects of social life. People are living their lives differently and experiencing space differently. Even recent ideas like that of virtual space have become inadequate very quickly, as new and complex forms of human technological interactions develop. The use of Twitter has become routine, changing the nature of politics, political action, news, social protest, marketing and all kinds of practices. The nature of the Arab Spring uprisings across the Middle East in 2011, which made heavy use of new web media, would have been unimaginable only a few years earlier. Research across the many sub-disciplines of human geography is becoming ever more preoccupied with these transformations, and geographical thinking offers a potentially powerful means to better understand the implications and impacts of these technological changes on today world.

These are just a limited number of the current research areas in human geography. More could be added, but that is beyond the scope of what this book is about. Its task has been to provide a crash course in human geography. At times this has undoubtedly meant doing the subject some injustice, either by leaving important topics out or by simplifying complex ideas. And in no sense has this overview of human geography been exhaustive. Rather, the book is meant to have served as a starting point for you, the reader. Its main aim is to give you an overall taste of this diverse social science subject, and inspire you to study it further. I hope you will discover how rewarding that can be.

GLOSSARY

Actor-network theory (ANT): A theoretical approach that sees all things in the world (in abstract terms) as made up of many different connections (translations, associations, mediations) to other things (people, objects, etc.). ANT questions key assumptions in Western knowledge about the conceptual boundaries between categories: humans/objects; nature/culture; tradition/ modernity. It argues that such conceptual divisions stop us seeing the world as it really is: a collection of diverse interconnected things in a constant state of being created.

Annihilation of space by time: The idea, associated with the work of Karl Marx, that the circulation of capital makes time the fundamental dimension of social life in its quest for profit. Space is 'annihilated' because the distance between, for example, markets for commodities matters less than the time it takes to get those commodities to market. Under capitalism, space is thus used, created and dominated to fit into the strict temporal constraints that the profitable circulation of capital requires.

Alternative food networks: A response to conventional food chains dominated by supermarkets and transnational firms in which local food producers market foodstuffs through alternative outlets (such as farmers' markets or home delivery).

Avant-garde: Used in English to describe works or groups at the forefront of new innovation, particularly in art, culture or politics.

Baden-Württemberg: One of the 16 states in Germany, in the south-west of the country.

Birth rate: The number of childbirths per 1,000 people in a population each year.

Bonds: Certificates that represent money a government or a corporation has borrowed. Once issued they can be traded in a 'secondary market' like shares.

British Empire: All the territories governed by Britain from the beginning of western European colonialism in the 16th century to the early 20th century. The British Empire varied in size over this period but was largest during the 19th century, when its territory included all the major continental land masses. At its height, it was the largest empire to have existed, accounting for around 450 million people and a quarter of the planet's land area.

Buffer zone: In Mackinder's Heartland thesis, these consisted of states not allied to one of the Great Powers and often lying between them as neutral territories that were the object of diplomatic manipulation.

Classical social theories: Works of historical thinkers that led to modern social science subjects including sociology, economics, politics and human geography from the late 18th century. Key thinkers in the 19th century include Auguste Comte, Karl Marx, Emile Durkheim and (later) Max Weber.

Colonialism: In general terms, this refers to the establishment and maintenance of colony territories in one area by people from another area of the world. Historically, the period from the late 15th until the 20th century is described as the 'colonial period'.

Containment theory: A theory of foreign policy developed by the US from the mid-1940s that used military, economic and diplomatic means to prevent the spread of communism as a political system during the Cold War.

Communist Cuba: After a three-year rebellion, a revolutionary force established a communist government in 1959 on the Caribbean island of Cuba. Led for most of this time by the rebel leader Fidel Castro, Cuba remains a communist state outside of the global capitalist economy.

Corporate governance: The mechanisms by which firms are managed and run including the interaction of boards of directors, shareholders and external regulation.

Cultural turn: An intellectual shift that brought issues of culture to forefront of debates in human geography and other social science disciplines.

Cultural anthropology: A branch of anthropology that focuses on the cultural variation between human societies in different places around the world.

Death rate: The number of deaths per 1,000 people in a population per year.

Derivatives: In finance, a product which is derived from an underlying asset (such as a share), that has no value in itself but is essentially a contract between two parties that specifies a set of conditions under which a payment will be made.

Domino effect: The idea within **containment theory** that if, for example, one state becomes communist, then neighbouring states are also likely to follow suit.

Ecosystem: A biological environment consisting of all the living organisms in a particular area along with the physical components of that environment (such as air, water, soil, light).

Ecotourism: Tourism in fragile or pristine areas of the world where **ecosystems** are protected, which intends to have a low impact on environments. It is often small in scale and intended as an alternative to commercial tourism

Economic sociology: This interdisciplinary field of social science adopts a different approach to the positivist models of neoclassical economies. It is concerned with the social consequences of economic exchanges, the social meanings they involve and the social interactions they facilitate or obstruct.

Environmental degradation: Refers to any change to an environment that humans regard as destructive or deleterious (which, of course, makes it a subjective judgement). Examples would include ecosystem destruction, species extinction or the depletion of natural resources.

Embodied space: A concept that seeks to capture how human experience and consciousness takes on material and spatial form through the nature of our bodies' existence in the world.

Fascist/Fascism: A radical political ideology that has many diverse forms but is generally associated with authoritarian rule, strong nationalism and usually an exclusionary sense of identity based on imagined community, ancestry, culture and race.

Financialization: A process where the economy becomes increasingly focused on profit made through financial channels rather than on commodity production or trade. It refers to the growing importance of financial markets, institutions, motives and elites in society.

Flexibilization: A process generally applied to a set of different changes to the nature of production associated with, for example, in manufacturing, new kinds of adaptable machinery to enable different models of goods to be produced and the use of part-time, contract and temporary labour to allow companies to respond to rapidly changing markets for their goods.

French Revolution: Sometimes known as the first French Revolution (there were later ones), the bloody revolution of 1789 saw the end of absolute monarchy in France and the foundation of a French Republic.

G8: The group of seven major economies founded in France in 1975 consisted of Canada, Italy, (West) Germany, Japan, the UK and the US. Russia joined in 1997 to become the G8.

Gentrification: An urban process whereby old residential housing or other buildings are upgraded and renovated by new owners in an urban district. In residential areas, this often involves change in the demographic character of a district.

German Ruhr: An urban area in north-west Germany along the River Rhine containing several large cities that have experienced deindustrialization (such as Essen and Dortmund).

Global North: A more recent term for what was previously described as the wealther 'First World' or 'advanced economies' (including wealthier economies in other regions outside the northern hemisphere such as Japan and Australia).

Global South: A more recent term for what was previously described as the 'Third World' or 'developing economies'.

Global village: Closely associated with the writer Marshall McLuhan and his book *The Gutenberg Galaxy* (1962), the idea denotes how the world is becoming a small place through, in particular, technology-facilitated globalization.

Global commodity/global value chain: A network of labour and production processes whose end result is a finished product or commodity.

Greenhouse gas: A gas within the Earth's atmosphere that absorbs and emits within the thermal infrared range, and is responsible for a warming effect. Key gases include water vapour and carbon dioxide.

Gross Domestic Product: The market value of all goods and services produced by a country in a given period.

Guerrilla resistance: A form of resistance, usually military, where a small number of combatants – that often includes non-professional armed individuals – use military tactics (such as ambushes and raids) alongside surprise and mobility to resist a larger and less mobile traditional army or force.

(Persian) Gulf War: A UN-authorized war against Iraq between 1990 and 1991, led by the US and other allies in response to the invasion and annexation of Kuwait.

Hedonism: A school of thought that argues pleasure is the only intrinsic good in the world and therefore strives for it above all else.

Holocaust: Literally meaning 'catastrophe', this refers to the genocide of around 6 million European Jews and millions of other social groups in a systematic programme of state-sponsored murder by Nazi Germany in the Second World War.

Homogeneity: A state whereby things or places are increasingly similar until they become indistinguishable from each other.

Humanism: A worldview, philosophy and/or practice that focuses on human concerns and affirms some concept of human nature.

Hybrid/hybridity: The product of a combination of things that are distinct, originating from biology (for example, plants) but used in human geography in relation to meaning, culture, values, etc.

Industrial Revolution: A period in which fundamental changes occurred in agriculture, textile and metal manufacture, transportation, economic policies and the social structure. This transformation began in Britain in the 18th century, continuing into the 19th and spreading to Europe, North America, Japan and subsequently the rest of the world.

Informationalization: the process by which economies, cultures and societies have become increasingly oriented around knowledge and information as organising factors.

International Criminal Court: A permanent tribunal founded in 2002 by treaty charged with prosecuting individuals for war crimes, crimes against humanity and genocide. It is located in The Hague.

International Monetary Fund: Founded in 1945, this is an intergovernmental organization that aims to foster economic cooperation with a particular focus on policies that affect currency exchange rates and the balance of payments. Its stated aim is to foster economic stability and growth.

IRA: The name of several military organizations aimed at bringing about a united Irish nation-state. In the later 20th century, the Provisional IRA used both violence and political methods to seek this goal.

Korean War: Beginning in June 1950 with an armistice signed in 1953, a war between the UN-backed South Korea and Chinese-backed North Korea. The war resulted in the division of capitalist South Korea from communist North Korea along the 38th parallel of latitude.

Kyoto Protocol: Adopted in 1997, this is an international treaty aimed at limiting greenhouse gas emissions to a level that will prevent dangerous levels of human-induced global warming. By 2011, 191 countries had signed the treaty.

Living standards/standard of living: The level of well-being of an individual usually measured in economic terms by income or output per person and associated with (but not necessarily equivalent to) their quality of life.

Modernity: A state of society that is post-tradition and can be traced back historically to the 16th and 17th centuries, associated with the emergence of capitalism, industrialization, rational scientific thought and the devlopment of nation-states and their associated institutions.

Mutually assured destruction (MAD): An aspect of military ideology that is based on the idea that two opposing forces using weapons of mass destruction (i.e. nuclear weapons) will lead to the total annihilation of both attacker and defender, thus acting as a deterrent to conflict.

New social movements: A range of issue-based movements that have emerged in Western societies since the 1960s and that exist within civil society rather than in the formal political institutions of nation-states (such as environmental or civil rights movements).

Non-representational theories: Attempts to theorize the social world as 'mobile practices' that focus on the potential of the flow of events in the moment rather than on static models of thought and action that dominate conventional social science.

Organisation for Economic Co-operation and Development (OECD): An international economic organization of 34 countries founded in 1961 with the aim of promoting economic growth and development.

Orient: A historical European representation of the Eastern world from the Middle East and encompassing all of Asia.

Outer rim: In Mackinder's Heartland thesis, states outside the core of the world geopolitical map including the USA and Japan.

Patriarchy/patriarchal structures: A system of social structures and practices through which men dominate, oppress and exploit women.

Phenomenology: A strand of continental European philosophy that centres on the significance of reflecting how the world can be understood through intellectual inquiry, especially in relation to the key role played by language.

Pivot area: In Mackinder's Heartland thesis, this was land locked central Asia including eastern Europe and was important as whoever controlled it would have the balance of global power.

Political economy: A diverse theoretical tradition that is critical of neoclassical economics and often (but not always) draws on Marxism to view the nature of economic activity as a politicised object of study.

Postcolonialism: A critical perspective concerned with the consequences of colonialism and its contestation on the peoples of both colonized and colonizing countries in the past, including representations and practices in the present.

Postmodern: Literally meaning 'after the modern', this a term that is used loosely to refer to an artistic movement, a state of society, and a condition of the philosophy of knowledge.

Poststructuralist: A term used to describe a group of French philosophers since the late 1970s who, while rejecting this label, all based their ideas on the common themes of rejecting a structuralist view of language and of an essentialized human subject.

Positivism: A philosophy of science that originated from Auguste Comte (1798–1857) that distinguishes it from metaphysics and religion through a prioritization of observation or accessible experience of the world (empiricism) and the construction of theories on that basis.

Positionality: The idea that where an individual is located in social structures and institutions affects how they understand the world.

Productivity: A measure of output relative to input which is usually expressed as the ratio of the returns from sales to the costs of production.

Psychoanalytic theories: founded in the work of Sigmund Freud (1856–1939), this broad range of theory is concerned with the nature of human subjectivity.

Qualitative methods: A set of research tools that seek to reveal how the world is viewed, experienced and constructed by social actors. It includes interviews, focus groups, participant observation and textual interpretation.

Quantitative methods: A set of research tools that use mathematical and statistical techniques to develop theories and proofs of social phenomena.

Silicon Valley: The southern part of the San Francisco Bay area in California, US, which has an historic concentration of many of the world's leading technology companies.

Socioeconomics: A broad term to describe theoretical approaches to understanding economic activity that involves social factors such as values, meanings and norms (as opposed to the approach of mainstream neoclassical economics).

Sovereign debt: Public or national debt that nation-state governments owe.

Spaceship Earth: An idea that became popular from the 1960s when the first images of Earth from space were taken. Often associated with the Green movement's worldview of Earth as a single, contained, living organic system bounded in space.

Spatiality: The socially produced nature of space.

Spin-off firms: New firms that are formed by former employees of an existing firm based on a new idea or aspect of the existing firm's activity.

Subjectivity: The property of human beings that leads to their sense of identity and understanding of what the world 'is'.

Surplus value: Developed significantly by Marx and in Marxist thinking, this idea refers to the additional value created by workers when they produce a good or service in excess of the cost of their own labour and other inputs.

Time-space convergence/compression: A decrease in what is known as the friction of (or barriers/problems created by) distance between places often associated with improvements in transport and communication. The related idea of 'convergence' adds to this an experiential aspect as we feel the world becoming smaller (and is linked to Marxist thinking).

Trade justice: A concept and civil society campaign to change the rules of global trade so that poorer countries and people benefit, based on the view that current trade reflects an unfair advantage to the richer and more powerful nations.

Total global output: The sum of all the goods and services produced by the global economy.

UN Security Council: One of the key bodies within the UN charged with maintaining international peace and security, and with the authority to authorize sanctions and military action.

Underdevelopment: A lack of development, usually applied to countries or regions, and based on the historical comparison with a subjective view of a greater degree of development.

Vietnam War: A Cold War military conflict between the US and non-communist allies and North Vietnamese communist groups and their allies. The war encompassed contemporary Vietnam, Laos and Cambodia and lasted from 1955 to 1975.

Western imperialism: The creation and reproduction of unequal relationships between Western states and others in Africa, Asia and the Americas that are based on domination and subordination.

World Bank: Created in 1944, this is an international financial institution charged with providing loans to developing countries to finance capital investment programmes.

REFERENCES

Adey, P. (2009) *Mobility: Key Ideas in Geography*. London: Routledge.

Anderson, K. and Smith, S. (2001) 'Editorial: Emotional geographies', *Transactions of the Institute of British Geographers* 26: 7–10.

Amin, A. and Cohendet, P. (2004) *Architectures of Knowledge: Firms, Capabilities, and Communities*. Oxford: Oxford University Press.

Barnes, T. (2008) 'Making space for the market: Live performances, dead objects, and economic geography', *Compass* 3: 1–17.

——(2011) 'Notes from the underground: Why the history of twentieth century Anglo-American economic geography matters', lecture presented at the Association of American Geographers Annual Conference, Seattle, WA.

Barnett, C. and Lowe, M. (2004) *Spaces of Democracy: Geographical Perspectives on Citizenship, Participation and Representation*. London: Sage.

Bathelt, H., Malmberg, A. and Maskell, P. (2004) 'Clusters and knowledge: Local buzz, global pipelines, and the process of knowledge creation', *Progress in Human Geography* 28(1): 31–56.

Betsill, M. M. and Bulkeley, H. (2005) *Cities and Climate Change: Urban Sustainability and Global Environmental Governance*. London; Routledge.

Berndt, C. and Boeckler, M. (2009) 'Geographies of circulation and exchange: Constructions of markets', *Progress in Human Geography* 33(4): 535–51.

Binnie, J. and Valentine, G. (1999) 'Geographies of sexuality: A review of progess', *Progress in Human Geography* 23: 175–87.

Binnie, J. and Skeggs, B. (2004) 'Cosmopolitan knowledge and the production and consumption of sexualized space: Manchester's gay village', *The Sociological Review* 52: 39–61.

Bonnett, A. (2008) 'Whiteness and the West', in C. Dwyer and C. Bressey (eds) New Geographies of Race and Racism. Aldershot: Ashgate: 17–28.
Browne, K. (2006) 'Challenging queer geographies', *Antipode* 38(5): 885–93.
——(2009) 'Naked and dirty: Rethinking (not) attending festivals', *Tourism and Cultural Change* 7(2): 118–32.
Browne, K., Lim, J. and Brown, G. (eds) (2009) *Geographies of Sexuality*. Aldershot: Ashgate.
Bryman, A. (1995) *Disney and His Worlds*. London: Routledge.
Bryson, J. and Henry, N. (2005) 'The global production system: From Fordism to post-Fordism', in J. Sidaway, P. Daniels, M. Bradshaw and D. Shaw (eds) *Human Geography: Issues for the 21st Century*. Harlow: Prentice Hall.
Bulkeley, H. (2005) 'Reconfiguring environmental governance: Towards a politics of scales and networks', *Political Geography* 24(8): 875–902.
Butler, J. (1990) *Gender Trouble: Feminism and the Subversion of Identity*. London: Routledge.
——(1993) *Bodies that Matter: On the Discursive Limits of Sex*. London: Routledge.
Callon, M. (1998) *The Laws of the Markets*. Oxford: Blackwell.
Castles, S. and Miller, J. (2003) *The Age of Migration: International Population Movements in the Modern World*. Basingstoke: Palgrave Macmillan.
Castells, M. (2009) *The Rise of the Network Society* [2nd edition]. Oxford: Blackwell.
Castree, N., Coe, N., Ward, K. and Samers, M. (2004) *Spaces of Work*. London: Sage.
Chatterji, J. (2007) *The Spoils of Partition: Bengal and India 1947–67*. Cambridge: Cambridge University Press.
Church, A. and Coles, T. (eds) (2006) *Tourism, Power and Space*. London: Routledge.
Cloke, P. (2000) 'Rural', in R. Johnston, D. Gregory, G. Pratt and M. Watts (eds) *The Dictionary of Human Geography* [4th edition]. Oxford: Blackwell
——(2005a) 'Conceptualising rurality', in P. Cloke, T. Marsden and P. Mooney (eds) *Handbook of Rural Studies*. London: Sage.
——(2005b) 'Self-other', in P. Cloke, P. Crang and M. Goodwin (eds) *Introducing Human Geographies* [2nd edition]. Arnold: London.
——(2009) 'Rural', in R. Johnson, D. Gregory, G. Pratt, M. Watts and S. Whatmore (eds) *The Dictionary of Human Geography*. Oxford: Blackwell.
Cochrane, A. (2008) 'Cities: Urban worlds', in P. Daniels, M. Bradshaw and D. Shaw (eds) *An Introduction to Human Geography* [3rd edition]. Harlow: Prentice Hall.
Corrigan, P. (1997) *The Sociology of Consumption: An Introduction*. London: Sage.
Crang, M. (1998) *Cultural Geography*. London: Routledge.

Crang, P. and Jackson, P. (2001) 'Comsuming geographies', in K. Morley and K. Robins (eds) *British Cultural Studies*. Oxford: Oxford University Press.

Crang, P., Dwyer, C. and Jackson, P. (2003) 'Transnationalism and the spaces of commodity culture', *Progress in Human Geography* 27: 438–56.

Desforges, L. (2005) 'Travel and tourism', in P. Cloke, P. Crang and M. Goodwin (eds) *Introducing Human Geographies* [2nd edition]. Arnold: London.

Dicken, P. (2011) *Global Shift: Mapping the Changing Contours of the Global Economy* [6th edition]. London: Sage.

Driver, F. (2008) 'Imaginative geographies', in P. Cloke, P. Crang and M. Goodwin (eds) *Introducing Human Geographies* [2nd edition]. Arnold: London.

Escobar, A. (1995) *Encountering Development: The Making and Unmaking of the Third World*. Princeton: Princeton University Press.

Faulconbridge, J. (2008) 'Negotiating cultures of work in transnational law firms', *Journal of Economic Geography* 8: 497–517.

Florida, R. (2002) *The Rise of the Creative Class*. New York: Basic Books.

Friedman, T. (2007) *The World Is Flat: The Globalized World in the Twenty-First Century* [2nd edition]. New York: Penguin.

Gibson-Graham, J. K. (1996) *The End of Capitalism (As We Knew It): A Feminist Critique of Political Economy*. Minneapolis: University of Minnesota Press.

——(2008) 'Diverse economies: Performative practices for "other worlds"', *Progress in Human Geography* 32(5): 613–32.

Glückler, J. (2005) 'Making embeddedness work: Social practice institutions in foreign consulting markets', *Environment and Planning A* 37(10): 1727–50.

Grabher, G. (2001) 'Ecologies of creativity: The village, the group, and the heterarchic organization of the British advertising industry', *Environment & Planning A* 33(2): 351–74.

Grundy-Warr, K. and Sidway, J. (2008) 'The place of the nation state', in P. Daniels, M. Bradshaw and D. Shaw (eds) *An Introduction to Human Geography* [3rd edition]. Harlow: Prentice Hall.

Hamnett, C. (2003) *Unequal City: London in the Global Arena*. London; Routledge.

Harvey, D. (1989) *The Condition of Postmodernity*. Oxford: Blackwell.

——(2007) *The Limits to Capital* [2nd Edn]. London: Verso.

——(2011) *The Enigma of Capital and the Crises of Capitalism*. London: Profile Books.

Holloway, S. and Valentine, G. (2000) *Children's Geographies: Playing, Living, Learning*. London: Routledge.

Howell, P. (2005) 'Prostitution and the place of empire: Regulation and repeal in Hong Kong and the British imperial network', in L. Proudfoot and M. Roche (eds) *(Dis)placing Empire: Renegotiating British Colonial Geographies*. Aldershot, Ashgate: 175–97.

Hubbard, P. (2004) 'Cleansing the metropolis: Sex work and the politics of zero tolerance', *Urban Studies* 41(9): 1687–1702.

Ilbery, B. and Maye, D. (2008) 'Changing geographies of food production and consumption', in P. Daniels, M. Bradshaw and D. Shaw (eds) *An Introduction to Human Geography* [3rd edition]. Harlow: Prentice Hall.

Jacobs, J. (2010) *Sex, Tourism and then Post-Colonial Encounter: Landscapes of Longing in Egypt*. Aldershot: Ashgate.

James, A., Jenks, C. and Prout, A. (eds) (1998) *Theorizing Childhood*. Cambridge: Polity Press.

Jones, A. (2011) 'Theorising international youth volunteering: Training for global (corporate) work?' *Transactions of the Institute of British Geographers 36*, 4: 530–44.

Jones, A. and Murphy, J. (2010) 'Theorizing practice in economic geography: Foundations, challenges,and possibilities', *Progress in Human Geography* 35(3): 366–92.

Klein, N. (2000) *No Logo: Solutions for a Sold Planet*. London: Flamingo.

Larner, W. (2005) 'Neoliberalism in (regional) theory and practice: The stronger communities action fund in New Zealand', *Geographical Research* 43(1): 9–18.

Law, J. (1993) *Organising Modernity: Social Order and Social Theory*. Oxford: Blackwell.

Latour, B. (2005) *Reassembling the Social: An Introduction to Actor-Network Theory*. Oxford: Oxford University Press.

Lee, R. (2006) 'The ordinary economy: Tangled up in values and geography', *Transactions of the Institute of British Geographers* 31(4): 413–32.

Little, J. (2003) 'Riding the rural love train: Heterosexuality and the rural community', *Sociologia Ruralis* 4: 401–17.

Longhurst, R. (1997) '(Dis)embodied geographies', *Progress in Human Geography* 21: 486–501.

Lovelock, J. (2000) The *Ages of Gaia: The Biography of Our Living Earth*. London: Penguin.

Mackenzie, D. (2008) *Material Markets: How Economic Agents Are Constructed*. Oxford: Oxford University Press.

Malmberg, A. and Maskell, P. (2002) 'The elusive concept of localisation economies: Towards a knowledge-based theory of spatial clustering', *Environnent & Planning A* 34: 429–49.

Maskell, P., Eskelinen, H., Hannibalsson, I., Malmberg, A. and Vatne, E. (1998) *Competitiveness, Localised Learning and Regional Development: Specialisation and Prosperity in Small Open Economies*. London: Routledge.

Matless, D. (1995) 'The art of right living: Landscape and citizenship 1918–39', in S. Pile and N. Thrift (eds) *Mapping the Subject: Geographies of Cultural Transformation*. London: Routledge.

McCann, E. and Ward, K. (2011) *Mobile Urbanism: Cities and Policies in a Global Age*. Minneapolis: University of Minnesota Press.

McDowell, L. (1997) *Capital Culture: Gender at Work in the City of London*. Oxford: Blackwell.

Monk, J. and Hanson, S. (1982) 'On not excluding half the human in human geography', *The Professional Geographer* 34: 11–23.

Moss, P. (1999) 'Autobiographical notes on chronic illness', in R. Butler and H. Parr (eds) *Mind and Body Spaces: Geographies of Illness, Impairment and Disability*. London: Routledge.

Mowforth, M. and Munt, I. (2008) *Tourism and Sustainability: Development, Globalization and New Tourism in the Third World* [3rd edition]. London: Routledge.

Muir, R. (1997) *Political Geography*. Basingstoke: Macmillan.

Nayak, A. (2003) 'Last of the "Real Geordies"? White masculinities and the subcultural response to deindustrialisation', Environment and Planning D: Society and Space 21(1): 7–25.

Ohmae, K. (1996) *The End of the Nation State: The Rise of Regional Economies*. London: Penguin.

Painter, J. and Philo, C. (eds) (1995) 'Spaces of citizenship', *Political Geography* 14 Special Issue.

Parker, G. (2002) *Citizenships, Contingency and the Countryside*. London: Routledge.

Parr, H. (2005) 'Emotional geographies', in P. Cloke, P. Crang and M. Goodwin (eds) *Introducing Human Geographies* [2nd edition]. Arnold: London.

Perrons, D. (1995) 'Gender inequalities in regional development', *Regional Studies* 29(5): 465–76.

Philo, C. (1992) 'Neglected rural geographies: A review', *Journal of Rural Studies* 8: 193–207.

——(1997) 'Of other rurals', in P. Cloke and J. Little (eds) *Contested Countryside Cultures: Otherness, Marginality and Rurality*. London: Routledge.

Pile, S. (2010) 'Emotions and affect in recent human geography', *Transactions of the Institute of British Geographers* 35(5): 5–20.

Porter, M. (1998) 'Clusters and the new economics of competition', *Harvard Business Review* December: 77–90.

Pratt, G. (2005) 'Masculinity – femininity', in P. Cloke, P. Crang, and M. Goodwin (eds) *Introducing Human Geographies* [2nd edition]. Arnold: London.

Punch, S. (2001) 'Household division of labour: Generation, age, gender, birth order and sibling composition', *Work, Employment and Society* 15: 803–23.

Rennie Short, J. (2004) *Global Metropolitan*. London: Routledge.

Roberts, P. and Sykes, H. (1999) *Urban Regeneration: A Handbook*. London: Sage.

Rose, G. (1993) *Feminisim and Geography: The Limits to Geographical Knowledge*. Cambridge: Polity.
Said, E. (1979) *Orientalism*. London: Penguin.
Sassen, S. (2001) *The Global City* [2nd edition]. Princeton: Princeton University Press.
Shaw I. and Warf, B. (2009) 'Worlds of affect: Virtual geographies of video games', *Environment and Planning A* 41(6): 1332–43.
Sibley, D. (1999) 'Creating geographies of difference', in D. Massey, J. Allen and P. Sarre (eds) *Human Geography Today*. Cambridge: Polity.
Sidaway, J. (2008) 'Geopolitical traditions', in P. Daniels, M. Bradshaw and D. Shaw (eds) *An Introduction to Human Geography* [3rd edition]. Harlow: Prentice Hall.
Smith, S. (2009) 'Citizenship in Johnson', in R. J. Johnston, D. Gregory, G. Pratt, M. Watts and S. Whatmore (eds) *The Dictionary of Human Geography*. Oxford: Blackwell.
Soja, E. (2000) *Postmetropolis: Critical Studies of Cities and Regions*. London: Verso.
Storper, M. (1995) 'The resurgence of regional economies, ten years later: The region as a nexus of untraded interdependencies', *European Urban and Regional Studies* 2: 191–221.
——(1997) *The Regional World: Territorial Development in a Global Economy*. London: Guilford Press.
Storper, M. and Venables, A. (2004) 'Buzz: Face-to-face contact and the urban economy', *Journal of Economic Geography* 4: 351–70.
Taylor, P. (2004) *World City Network: A Global Urban Analysis*. London: Routledge.
Thrift, N. (1996) *Spatial Formations*. London: Sage.
——(1997) 'The still point: Expressive embodiment and dance', in S. Pile and M. Keith (eds) *Geographies of Resistance*. London: Routledge: 124–51.
Thrift, N. and Pile, S. (1995) *Mapping the Subject: Geographies of Cultural Transformation*. London: Routledge.
Turner, B. (1992) *Regulating Bodies: Essays in Medical Sociology*. (London: Routledge)
UN AIDS (2010) *UN Aids Report of the Global Academic*. Downloadable at www.unaids.org.
UNHCR (2010) *UNHCR Global Report*. Geneva: UNHCR.
Valentine, G. (1993) '(Hetero)sexing space: Lesbian perceptions and experiences of everyday spaces', *Environment and Planning D: Society and Space* 11: 395–413.
——(1995) 'Out and about: A geography of lesbian communities', *International Journal of Urban and Regional Research* 19: 96–111.
——(2001) *Social Geographies Space and Society*. Harlow: Prentice Hall.
——(2004) 'Moral geographies of sexual citizenship' in R. Lee and D. Smith (eds) *Geographies and Moralities: International Perspectives on Development, Justice and Place*. Oxford: Blackwell.

Vanderbeck, R, Andersson, J., Valentine, G., Ward, K. and Sadgrove, J. (2011) 'Sexuality, activism, and witness in the Anglican Communion: The 2008 Lambeth Conference of Anglican Bishops', *Annals of the Association of American Geographers*.

Williams, S. (2009) *Tourism Geography: A New Synthesis*. London: Routledge.

Winchester, H. and White, P. (1988) 'The location of marginalised groups in the inner city', *Environment & Planning D: Society and Space* 6: 37–54.

Wylie, J. (2007) *Landscape*. London: Routledge.

Young, I. (1990) *Justice and the Politics of Difference*. Princeton: Princeton University Press.

INDEX